THE MYTH OF JUNK DNA

THE MYTH OF JUNK DNA

JONATHAN WELLS

SEATTLE   DISCOVERY INSTITUTE PRESS   2011

Description

According to a number of leading proponents of Darwin's theory, "junk DNA"—the non-protein coding portion of DNA—provides decisive evidence for Darwinian evolution and against intelligent design, since an intelligent designer would presumably not have filled our genome with so much garbage. But in this provocative book, biologist Jonathan Wells exposes the claim that most of the genome is little more than junk as an anti-scientific myth that ignores the evidence, impedes research, and is based more on theological speculation than good science.

Copyright Notice

Publisher's Note
This book is part of a series published by the Center for Science & Culture at Discovery Institute in Seattle. Previous books include *The Deniable Darwin* by David Berlinski, *In the Beginning and Other Essays on Intelligent Design* by Granville Sewell, *God and Evolution: Protestants, Catholics, and Jews Explore Darwin's Challenge to Faith*, edited by Jay Richards, and *Darwin's Conservatives: The Misguided Quest* by John G. West.

Library Cataloging Data
The Myth of Junk DNA by Jonathan Wells (1942–)

Illustrations by Ray Braun

174 pages, 6 x 9 x 0.4 inches & 0.6 lb, 229 x 152 x 10 mm. & 0.26 kg

Library of Congress Control Number: 2011925471

BISAC: SCI029000 SCIENCE / Life Sciences / Genetics & Genomics

BISAC: SCI027000 SCIENCE / Life Sciences / Evolution

ISBN-13: 978-1-9365990-0-4 (paperback)

Publisher Information
Discovery Institute Press, 208 Columbia Street, Seattle, WA 98104

Internet: http://www. discoveryinstitutepress.com/

Published in the United States of America on acid-free paper.

First Edition, First Printing. May 2011.¡™¢

Praise for *The Myth of Junk DNA*

"JONATHAN WELLS HAS CLEARLY DONE HIS HOMEWORK. IN *THE MYTH of Junk DNA*, he cites hundreds of research articles as he describes the expanding story of non-coding DNA—the supposed 'junk DNA.' It is quite possibly the most thorough review of the subject available. Dr. Wells makes it clear that our early understanding of DNA was incomplete, and genomics research is now revealing levels of control and complexity inside our cells that were undreamed of in the 1980s. Far from providing evidence for Darwinism, the story of non-coding DNA rather serves to increase our appreciation for the design of life."

Ralph Seelke, Ph.D.
Professor of Microbial Genetics and Cell Biology
University of Wisconsin-Superior

"CITING HUNDREDS OF PEER-REVIEWED ARTICLES WHICH SHOW THAT more and more of the genome is functional, Jonathan Wells delivers a powerful and carefully researched broadside against the 'junk DNA hypothesis.' Even biologists who firmly reject the notion of intelligent design must surely acknowledge on the evidence presented in this timely book that appealing to 'junk DNA' to defend the Darwinian framework no longer makes any sense."

Michael Denton, Ph.D.
Medical Geneticist and Author of Nature's Destiny

"THIS IS AN EXCELLENT AND IN-DEPTH DISCUSSION OF SEVERAL KEY points of the subject of 'junk-DNA.' The author shows for many prime examples still advanced by leading neo-Darwinians that the 'Darwin-of-the-gaps' approach doesn't function or is at least doubtful."

Wolf-Ekkehard Lönnig, Ph.D.
Senior Scientist, Department of Molecular Plant Genetics
Max Planck Institute for Plant Breeding Research (retired)

"THERE IS A BOX IN THE BIOLOGICAL SCIENCES INTO WHICH ALL EVI-dence must be placed. That box is called Darwinian evolution. In *The Myth of Junk DNA* Jonathan Wells tells the intriguing story of 'junk' DNA—the idea that non-protein coding DNA, which accounts for the majority of the DNA in the genome, is non-functional and without purpose; the result of the unguided purposeless process of random mutation and natural selection that produced it. In recent years, however, numerous researchers—not necessarily opponents of Darwinian evolution or advocates of intelligent design—have discovered many functions for non-protein coding DNA, which are thoroughly reviewed by Wells in this book. Unfortunately, in their effort to keep the 'junk' label attached to non-protein coding DNA so that it remains in the box of Darwinian evolution, a number of prominent Darwinists continue to insist, in spite of the recent results to the contrary, that it is largely left-over waste from the evolutionary process. As Wells clearly demonstrates in his book, this dogmatic commitment inhibits the scientific process. Science needs to be guided by objective evaluation of the evidence, and scientists should not allow their thinking to be arbitrarily restricted by dogmatic ideas. We need scientists who think outside the Darwinian box. Wells's book not only informs its readers of very recent research results, but also encourages them to think objectively and clearly about a key discovery in biology and to approach biological research with more creativity. It is a great read."

Russell W. Carlson, Ph.D.
Professor of Biochemistry and Molecular Biology
University of Georgia

"FOR YEARS, DARWINISTS HAVE CLAIMED THAT MOST DNA IS LEFT-OVER detritus from failed evolutionary experiments. This 'junk DNA' has been offered as evidence for Darwinism and evidence against intelligent design. The only problem with the claim, as Jonathan Wells shows in this fascinating book, is that it's not true. Careful scientists have known for some time that the non-coding regions of DNA have all manner of function, so it is surprising to see prominent Darwinian scientists and their spokesmen continue to push the party line. Now that the evidence against the junk DNA story is indisputable, its defenders will want to beat a hasty retreat. *The Myth of Junk DNA* will make it hard for them to cover their tracks."

Jay Richards, Ph.D.
Co-Author, The Privileged Planet, and Editor, God and Evolution

Contents

ILLUSTRATIONS

PREFACE

THE DISCOVERY IN THE 1970S THAT ONLY A TINY PERCENTAGE OF our DNA codes for proteins prompted some prominent biologists at the time to suggest that most of our DNA is functionless junk. Although other biologists predicted that non-protein-coding DNA would turn out to be functional, the idea that most of our DNA is junk became the dominant view among biologists.

That view has turned out to be spectacularly wrong.

Since 1990—and especially after completion of the Human Genome Project in 2003—many hundreds of articles have appeared in the scientific literature documenting the various functions of non-protein-coding DNA, and more are being published every week.

Ironically, even after evidence for the functionality of non-protein-coding DNA began flooding into the scientific literature, some leading apologists for Darwinian evolution ratcheted up claims that "junk DNA" provides evidence for their theory and evidence against intelligent design. Since 2004, biologists Richard Dawkins, Douglas Futuyma, Kenneth Miller, Jerry Coyne and John Avise have published books using this argument. So have philosopher of science Philip Kitcher and historian of science Michael Shermer. So has Francis Collins, former head of the Human Genome Project and present director of the National Institutes of Health, despite the fact that he co-authored some of the scientific articles providing evidence against "junk DNA."

These authors claim to speak for "science," but they have actually been promoting an anti-scientific myth that ignores the evidence and relies on theological speculations instead. For the sake of science, it's time to expose the myth for what it is.

Far from consisting mainly of junk that provides evidence against intelligent design, our genome is increasingly revealing itself to be a multidimensional, integrated system in which non-protein-coding DNA performs a wide variety of functions. If anything, it provides evidence for intelligent design. Even apart from possible implications for intelligent design, however, the demise of the myth of junk DNA promises to stimulate more research into the mysteries of the genome. These are exciting times for scientists willing to follow the evidence wherever it leads.

I have tried to make this book as non-technical as possible, but some technical details are needed to make the case. To make things easier for non-biologists, I have included a glossary of basic technical terms at the end, and Chapter 9 contains brief summaries of the preceding chapters. Since the vitamin C pseudogene story is a detour from the main argument, I have omitted it from the main text but added it as an appendix.

My friends and colleagues Richard Sternberg and Paul Nelson have helped me enormously, though if this book contains errors they are mine alone. I am also grateful to my wife Lucy and my colleagues John West, Jay Richards and Casey Luskin for helping me to make the book more readable, to Ray Braun for doing the illustrations, and to the Center for Science & Culture at the Discovery Institute for its encouragement and financial support.

Seattle, 2011

1.

THE CONTROVERSY OVER
DARWINIAN EVOLUTION

WHY IS DARWINIAN EVOLUTION STILL SO CONTROVERSIAL? According to its defenders, there hasn't been any scientific controversy about it for years: The evidence for the theory is supposedly so overwhelming that it can now be regarded as a scientific fact.

Of course, if evolution meant only change over time, or minor changes within existing species, there would be no controversy. No sane person doubts the fact of change over time. And, indeed, there *is* overwhelming evidence for changes within existing species. Breeders have been observing or producing them for centuries.

But Darwinian evolution means much more than changes within existing species. Charles Darwin did not write a book titled *How Existing Species Change Over Time*; he wrote a book titled *The Origin of Species by Means of Natural Selection*. In fact, he argued that all living things are descendants of common ancestors that have been modified by unguided processes such as random variation and natural selection. (In the modern version of his theory—neo-Darwinism—variations are due to differences in genes, and new variations originate in genetic mutations.) According to Darwin, the same processes we now observe within species, if given enough time, produce new species, organs, and body plans.

Nevertheless, in 1937—almost eighty years after Darwin published *The Origin of Species*—neo-Darwinist Theodosius Dobzhansky noted that there was as yet no hard evidence to connect small-scale changes within existing species (which Dobzhansky called "microevolution") to the origin of new species or the large-scale changes we see in the fossil record (which he called "macroevolution"). But since "there is no way toward an understanding of the mechanisms of macroevolutionary changes, which require time on a geological scale, other than through a full comprehension of the microevolutionary processes observable within the span of a human lifetime," Dobzhansky concluded, "we are compelled at the present level of knowledge reluctantly to put a sign of equality between the mechanisms of macro- and microevolution, and proceeding on this assumption, to push our investigations as far ahead as this working hypothesis will permit."[1]

Sixty years after Dobzhansky wrote this, biologists had still not observed the origin of a new species ("speciation") by natural selection. In 1997, evolutionary biologist Keith Stewart Thomson wrote: "A matter of unfinished business for biologists is the identification of evolution's smoking gun," and "the smoking gun of evolution is speciation, not local adaptation and differentiation of populations."[2]

British bacteriologist Alan H. Linton looked for evidence of speciation and concluded in 2001: "None exists in the literature claiming that one species has been shown to evolve into another. Bacteria, the simplest form of independent life, are ideal for this kind of study, with generation times of twenty to thirty minutes, and populations achieved after eighteen hours. But throughout 150 years of the science of bacteriology, there is no evidence that one species of bacteria has changed into another... Since there is no evidence for species changes between the simplest forms of unicellular life, it is not surprising that there is no evidence for evolution from prokaryotic [e.g., bacterial] to eukaryotic [e.g., plant and animal] cells, let alone throughout the whole array of higher multicellular organisms."[3]

Of course, even if scientists eventually observe the origin of a new species by natural selection, the observation would not mean that natural selection can also explain the origin of significantly new organs or body plans. But the fact that scientists have not observed even the first step in macroevolution means that "evolution's smoking gun" is still missing.

Despite the lack of direct evidence for speciation by natural selection,[4] Darwin's followers still assume that he was essentially correct and regard changes within existing species as evidence for their theory. Thus generations of biology students have been taught about a shift in the relative proportions of light- and dark-colored peppered moths during the industrial revolution, about an increase in the proportion of large-beaked finches after a drought on the Galápagos Islands, and about the spread of antibiotic resistance among disease-causing bacteria. Indeed, pictures of peppered moths and Galápagos finches are so common in biology textbooks that I have called them "icons of evolution."[5]

Darwin believed that all living things are related in a "great Tree of Life," with the universal common ancestor at the base of the trunk and modern species at the tips of the branches.[6] Like peppered moths and the Galápagos finches, Darwin's Tree of Life is an icon of evolution, appearing in most modern biology textbooks.

Yet the evidence for Darwin's Tree of Life is far from overwhelming. The fossil record is fragmentary, and one of its most prominent features—the geologically abrupt appearance of major animal body plans in the Cambrian Explosion—contradicts Darwin's theory that major differences should arise only after millions of years of evolution, during which "the number of intermediate and transitional links" would have been "inconceivably great."[7] Darwin himself considered the absence of such links a serious problem, and subsequent fossil discoveries have aggravated it.[8-10]

Modern biologists have tried to overcome the problem by reconstructing evolutionary histories with comparisons of molecules in living species, but the molecular evidence is plagued with inconsistencies. Analyses of different molecules—or even the same molecule analyzed by two different laboratories—can yield different evolutionary trees. Indeed, molecular analyses have now persuaded even some evolutionary biologists to reject the hypothesis of a universal common ancestor.[11-13]

Many biology textbooks use drawings of the bones in vertebrate limbs (yet another icon of evolution) to illustrate "homology"—similarity of structure and position—which according to Darwin provides evidence for common ancestry. But most biologists before Darwin regarded homology as a result of common design. To establish that homology is due to common ancestry, neo-Darwinists have tried to explain it by the inheritance of similar genes, but developmental biologists know that this is not generally true.[14] Darwin's followers also tried to finesse the problem by re-defining homology to mean similarity due to common ancestry—but this means that homology can no longer be used as "evidence" for common ancestry without arguing in a circle: Similarity due to common ancestry is due to common ancestry.[15]

Darwin himself thought that the best evidence for his Tree of Life came from embryology, which he considered "by far the strongest single class of facts" in his favor.[16] He believed that vertebrate embryos are most similar in their earliest stages and become dissimilar as they develop, and that early embryos resemble the common ancestor of the whole group. German Darwinist Ernst Haeckel made drawings to illustrate this belief, and although his contemporaries pointed out that he had misrepresented the evidence, Haeckel's embryo drawings—another icon of evolution—were reprinted in biology textbooks for over a century. The truth is that vertebrate embryos start out looking very different from each other, then they converge somewhat in appearance midway through development before diverging again as they mature.[17–19]

So microevolution is a fact, supported by overwhelming evidence, but macroevolution remains an assumption, illustrated with icons that misrepresent the evidence or rely on circular reasoning. The icons are not science, but myth.

This may be one reason why—despite the Darwinists' near-monopoly over science education—most Americans still reject the doctrine that human beings evolved from ape-like ancestors by unguided processes such as random variation and survival of the fittest. To complicate matters, Darwin's defenders now face a new adversary: intelligent design.

According to intelligent design (ID), it is possible to infer from evidence in nature that some features of the world, and of living things, are better explained by an intelligent cause than by unguided natural processes. ID does not imply that design must be optimal or perfect;

indeed, as human artifacts show, something can be designed and yet be far from perfect. Unlike creationism, ID is not based on the Bible, but on evidence and logic; and unlike natural theology, ID does not argue for the existence of an omnipotent God (though it is consistent with God's existence). Nevertheless, Darwinists try to discredit ID as a form of religious fundamentalism—though their real objection is that it contradicts the Darwinian view that all features of living things can be explained by unguided natural processes.

So the old icons of evolution have failed to persuade most people that Darwinism is true, and intelligent design presents it with a new challenge. Accordingly, some of Darwin's defenders have turned to "junk DNA" to support their theory and refute ID.

In the 1950s, neo-Darwinists equated genes with DNA sequences and assumed that their biological significance lay in the proteins they encoded. But when molecular biologists discovered in the 1970s that most of our DNA does *not* code for proteins, neo-Darwinists called non-protein-coding DNA "junk" and attributed it to molecular accidents that have accumulated in the course of evolution. Like peppered moths, Galápagos finches, Darwin's Tree of Life, homology in vertebrate limbs, and Haeckel's embryos, "junk DNA" has become an icon of evolution. But is it science, or myth?

2.

JUNK DNA: THE LAST ICON OF EVOLUTION?

ONE SATURDAY MORNING IN 1953, AT THE CAVENDISH LABORA-tory in Cambridge, England, James Watson and Francis Crick con-cluded months of work by deciphering the molecular structure of deoxy-ribonucleic acid (DNA). They went to celebrate over drinks at a nearby pub, where Crick announced: "We have discovered the secret of life!"[1]

A century earlier, Charles Darwin had proposed his theory of evolu-tion by natural selection to explain how all living things are descended with modification from a common ancestor. Darwin's theory conflicted with the traditional and widespread notion that living things were de-signed. "There seems to be no more design in the variability of organic beings, and in the action of natural selection," Darwin wrote, "than in the course which the wind blows."[2] Although "I cannot look at the universe as the result of blind chance," he explained, "yet I can see no evidence of beneficent design, or indeed of design of any kind, in the details."[3] So he was "inclined to look at everything as resulting from designed laws, with the details, whether good or bad, left to the working out of what we may call chance."[4]

But Darwin did not know how traits are passed from generation to generation, much less how new traits originate. His contemporary Gregor Mendel performed experiments showing that several features of pea plants are determined by discrete factors that are inherited accord-ing to a few simple rules. (The factors were later named "genes" by Dan-ish botanist Wilhelm Johannsen.) Mendel found Darwin's theory un-persuasive, and Darwinists ignored his ideas for half a century.[5–6] It was not until the 1930s that Darwinian evolution and Mendelian genetics were combined in what became known as the neo-Darwinian synthesis.

According to neo-Darwinism, traits are passed on by genes that reside on microscopic thread-like structures in the cell called chromosomes, and new traits arise from accidental genetic mutations.

In the 1940s biochemists discovered that the active ingredient in chromosomes is DNA, and Watson and Crick's 1953 discovery that DNA consists of two complementary strands suggested a possible copying mechanism.[7-8] DNA consists of subunits called "nucleotides," each containing a sugar molecule attached to a phosphate group and one of four bases: adenine (A), thymine (T), cytosine (C) or guanine (G). In a DNA molecule, the nucleotides in each strand are attached by their phosphate groups, and the two strands wind around each other in a double helix. Since the A's and T's in one strand pair with T's and A's in the other, while the C's and G's pair with G's and C's, the nucleotide sequence in one strand is opposite and complementary to the sequence in the other strand. (**Figure 2.1**)

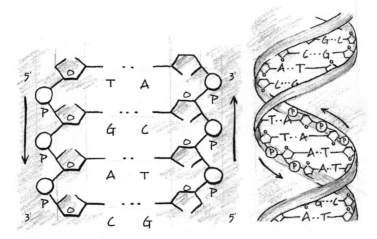

Figure 2.1 *The DNA double helix.* Idealized drawings of DNA in two dimensions (left) and three dimensions (right). Each nucleotide consists of a sugar group (pentagon) attached to a phosphate group (P) and one of four bases (A, T, C, G). The nucleotides are chemically connected only through their phosphate groups on the outside of the molecule. On the inside of the molecule the bases attract each other electrostatically, but because of their particular shapes the A's pair with T's and the C's pair with G's.

In 1958, Crick argued that "the main function of the genetic material" is to control the synthesis of proteins. According to Crick's "Sequence Hypothesis," the specificity of a segment of DNA "is expressed solely by the sequence of bases," and "this sequence is a (simple) code for the amino acid sequence of a particular protein." Crick further proposed that the sequence information in DNA is first transcribed into another molecule, ribonucleic acid (RNA), which is then translated into protein.[9]

As evidence for the copying mechanism and the process of protein synthesis accumulated, many biologists equated neo-Darwinian genes with DNA sequences. "With that," said French molecular biologist Jacques Monod in 1970, "and the understanding of the random physical basis of mutation that molecular biology has also provided, the mechanism of Darwinism is at last securely founded." As a consequence, Monod concluded, "Man has to understand that he is a mere accident."[10]

Following Monod's lead, and for the sake of simplicity, I will use "Darwinism" in the rest of this book to mean both Darwin's theory and neo-Darwinism.

In 1976, Oxford University professor and Darwinist Richard Dawkins wrote that the only "purpose" of DNA is to ensure its own survival. Dawkins considered the most important quality of successful genes to be "ruthless selfishness." It follows that "we, and all other animals, are machines created by our genes. Like successful Chicago gangsters, our genes have survived, in some cases for millions of years, in a highly competitive world." A body is simply "the genes' way of preserving the genes unaltered." Thus natural selection favors genes "which are good at building survival machines, genes which are skilled in the art of controlling embryonic development." And genes control embryonic development by encoding proteins.[11]

Junk DNA and Intelligent Design

YET BY 1970 biologists already knew that much of our DNA does not encode proteins. Although some suggested that non-protein-coding DNA might help to regulate the production of proteins from DNA templates, the dominant view was that non-protein-coding regions had no function.

In 1972, biologist Susumu Ohno (at the City of Hope National Medical Center in Los Angeles) published an article wondering why there is "so much 'junk' DNA in our genome."[12] The same year, his City of Hope colleague David Comings wrote that only about 20% of the human genome appears to be used; the remaining 80% seemed to be "junk"—though Comings did not necessarily think it was entirely useless.[13]

"The amount of DNA in organisms," Dawkins wrote in 1976, "is more than is strictly necessary for building them: A large fraction of the DNA is never translated into protein. From the point of view of the individual organism this seems paradoxical. If the 'purpose' of DNA is to supervise the building of bodies, it is surprising to find a large quantity of DNA which does no such thing. Biologists are racking their brains trying to think what useful task this apparently surplus DNA is doing. But from the point of view of the selfish genes themselves, there is no paradox. The true 'purpose' of DNA is to survive, no more and no less. The simplest way to explain the surplus DNA is to suppose that it is a parasite, or at best a harmless but useless passenger, hitching a ride in the survival machines created by the other DNA."[14]

In 1980, two papers appeared back to back in the journal *Nature*: "Selfish genes, the phenotype paradigm and genome evolution," by W. Ford Doolittle and Carmen Sapienza, and "Selfish DNA: The ultimate parasite," by Leslie Orgel and Francis Crick. The first paper argued that many organisms contain "DNAs whose only 'function' is survival within genomes," and that "the search for other explanations may prove, if not intellectually sterile, ultimately futile."[15] The second argued similarly that "much DNA in higher organisms is little better than junk," and its accumulation in the course of evolution "can be compared to the spread of a not-too-harmful parasite within its host." Since it is unlikely that such DNA has a function, Orgel and Crick concluded, "it would be folly in such cases to hunt obsessively for one."[16]

Two biologists wrote to *Nature* expressing their disagreement. Thomas Cavalier-Smith considered it "premature" to dismiss non-protein-coding DNA as junk,[17] and Gabriel Dover wrote that "we should not abandon all hope of arriving at an understanding of the manner in which some sequences might affect the biology of organisms in completely novel and somewhat unconventional ways."[18] So some biologists were skeptical of the notion of "junk DNA" from the very beginning—though most accepted it.

This does not mean that skeptics of "junk DNA" such as Cavalier-Smith and Dover were also skeptics of Darwinian evolution. In 1980, the most prominent opposition to Darwinism came from biblical creationists. A few years later, however, a new form of opposition appeared: intelligent design. In 1984, chemist Charles B. Thaxton, materials scientist Walter L. Bradley and geochemist Roger L. Olsen published *The Mystery of Life's Origin*, which criticized the idea that unguided natural processes produced the first living cells and which proposed that DNA had an intelligent cause at the beginning.[19] The following year, molecular biologist Michael Denton published *Evolution: A Theory in Crisis*, which critically analyzed the evidence for Darwin's theory and defended the view that design could be inferred from living things.[20]

In 1991, Berkeley law professor Phillip E. Johnson published *Darwin on Trial*, which concluded: "Darwinist scientists believe that the cosmos is a closed system of material causes and effects, and they believe that science must be able to provide a naturalistic explanation for the wonders of biology that appear to have been designed for a purpose. Without assuming those beliefs they could not deduce that common ancestors once existed for all the major groups of the biological world, or that random mutations and natural selection can substitute for an intelligent designer."[21]

In 1994, Brown University biologist (and co-author of some widely used high school biology textbooks) Kenneth R. Miller defended Darwinian evolution against the idea that living things are intelligently designed. He wrote: "The human genome is littered with pseudogenes, gene fragments, 'orphaned' genes, 'junk' DNA, and so many repeated copies of pointless DNA sequences that it cannot be attributed to anything that resembles intelligent design. If the DNA of a human being or any other organism resembled a carefully constructed computer program, with neatly arranged and logically structured modules each written to fulfill a specific function, the evidence of intelligent design would be overwhelming. In fact, the genome resembles nothing so much as a hodgepodge of borrowed, copied, mutated, and discarded sequences and commands that has been cobbled together by millions of years of trial and error against the relentless test of survival. It works, and it works brilliantly; not because of intelligent design, but because of the great blind power of natural selection to innovate, to test, and to discard what fails in favor of what succeeds." Indeed, Miller wrote, intelligent design theory "requires that we pretend to know less than we do about living organisms" and "requires a retreat back into an unknowledge of biology that is unworthy of the scientific spirit of this century."[22]

Using Junk DNA as Evidence for Darwinism and Against Intelligent Design

SEVERAL RECENT books have likewise used junk DNA as evidence for Darwinism and evidence against design or a creator. In 2004, Richard Dawkins wrote: "Genomes are littered with nonfunctional pseudogenes, faulty duplicates of functional genes that do nothing, while their functional cousins (the word doesn't even need scare quotes) get on with their business in a different part of the genome. And there's lots more DNA that doesn't even deserve the name pseudogene. It too is derived by duplication, but not duplication of functional genes. It consists of multiple copies of junk, 'tandem repeats', and other nonsense which may be useful for forensic detectives but which doesn't seem to be used in the body itself. Once again, creationists might spend some earnest time speculating on why the Creator should bother to litter genomes with untranslated pseudogenes and junk tandem repeat DNA."[23]

Biologist and textbook-writer Douglas J. Futuyma wrote in 2005 that the data enable us to "identify several patterns that confirm the historical reality of evolution." One of those patterns is that "every eukaryote's genome contains numerous nonfunctional DNA sequences, including pseudogenes: silent, nontranscribed sequences that retain some similarity to the functional genes from which they have been derived." Although Futuyma acknowledged that some "noncoding DNA is unlikely to be 'junk' (as was postulated in the early 1970s)," nevertheless only Darwinian evolution "can explain why the genome is full of 'fossil' genes: pseudogenes that have lost their function"—a phenomenon that is "hard to reconcile with beneficent intelligent design."[24]

In 2006, *Skeptic Magazine* publisher Michael Shermer wrote: "We have to wonder why the Intelligent Designer added to our genome junk DNA, repeated copies of useless DNA, orphan genes, gene fragments, tandem repeats, and pseudogenes, none of which are involved directly in the making of a human being. In fact, of the entire human genome, it appears that only a tiny percentage is actively involved in useful protein production. Rather than being intelligently designed, the human genome looks more and more like a mosaic of mutations, fragment copies, borrowed sequences, and discarded strings of DNA that were jerry-built over millions of years of evolution."[25]

The same year Francis S. Collins, former head of the Human Genome Project and now Director of the U.S. National Institutes of Health, wrote that "junk DNA" provides evidence for Darwin's theory of evolution. According to Collins, moveable segments of DNA known as "ancient repetitive elements" (AREs) have no function other than their own survival. "Some might argue," Collins wrote, "that these are actually functional elements placed there by the Creator for a good reason, and our discounting of them as 'junk DNA' just betrays our current level of ignorance. And indeed, some small fraction of them may play important regulatory roles. But certain examples severely strain the credulity of that explanation. The process of transposition often damages the jumping gene. There are AREs throughout the human and mouse genomes that were truncated when they landed, removing any possibility of their functioning. In many instances, one can identify a decapitated and utterly defunct ARE in parallel positions in the human and the mouse genome." This provides compelling support for Darwinian evolution, Collins argued, "unless one is willing to take the position that God has placed these decapitated AREs in these precise positions to confuse and mislead us."[26]

In 2007, Columbia University philosophy professor Philip Kitcher argued that "if you were designing the genomes of organisms, you would certainly not fill them up with junk." Yet "the most striking feature of the genome analyses we now have is how much apparently nonfunctional DNA there is." According to Kitcher, "From the Darwinian perspective all this is explicable—the molecular equivalent of tinkering that is pervasive in the history of life at the anatomical level… Over the history of life, the residues of past tinkering accumulate in the genome, the once-functional sequences, the degraded remains of genes, the long repeats." Junk DNA is also evidence against intelligent design (ID): "Why does Intelligence not eliminate the accumulations of junk and structures that have lost their original functions?" Kitcher argued that ID "would commit Intelligence to a whimsical tolerance of bungled designs."[27]

The following year, Kenneth R. Miller reaffirmed his view that pseudogenes provide evidence for Darwinian evolution and evidence against intelligent design. Humans lack a functional gene for an enzyme (abbreviated GLO) that is needed to synthesize vitamin C. As a result, we must include vitamin C in our diets, otherwise we suffer from scurvy. "But the interesting part of the story," Miller wrote, "is that we aren't exactly missing the *GLO* gene. In fact, it's right there on chromosome 8, in pretty much the same relative position in our genome where it is found in other mammals." (The names of genes are customarily italicized, while the names of their protein products are not.) Miller continued: "The problem is that our copy of the *GLO* gene has accumulated so many mutations, in the form of changes in the DNA base sequence, that it no longer works… If the designer wanted us to be dependent on vitamin C, why didn't he just leave out the *GLO* gene from the plan for our genome? Why is its corpse still there?" According to Miller, the presence of the *GLO* pseudogene is consistent with an evolutionary explanation but inconsistent with intelligent design.[28]

In 2009, University of Chicago geneticist Jerry A. Coyne compared predictions based on intelligent design with those based on Darwinian evolution. "If organisms were built from scratch by a designer," he argued, they would not have imperfections. "Perfect design would truly be the sign of a skilled and intelligent designer. *Imperfect* design is the mark of evolution; in fact, it's precisely what we *expect* from evolution." According to Coyne, "when a trait is no longer used, or becomes reduced, the genes that make it don't instantly disappear from the genome: Evolution stops their action by inactivating them, not snipping them out of the DNA. From this we can make a prediction. We expect to find, in the genomes of many species, silenced, or 'dead,' genes: genes that once were useful but are no longer intact or expressed. In other words, there should be vestigial genes." In contrast, creation by design predicts that no such genes would exist.

"Thirty years ago we couldn't test this prediction," Coyne continued, "because we had no way to read the DNA code. Now, however, it's quite easy to sequence the complete genome of species, and it's been done for many of them, including humans. This gives us a unique tool to study evolution when we realize that the normal function of a gene is to make a protein—a protein whose sequence of amino acids is determined by the sequence of nucleotide bases that make up the DNA. And once we have the DNA sequence of a given gene, we can usually tell if it is expressed normally—that is, whether it makes a functional protein—or whether it is silenced and makes nothing. We can see, for example, whether mutations have changed the gene so that a usable protein can no longer be made, or whether the 'control' regions responsible for turning on a gene have been inactivated. A gene that doesn't function is called a *pseudogene*."

According to Coyne, "the evolutionary prediction that we'll find pseudogenes has been fulfilled—amply. Virtually every species harbors dead genes, many of them still active in its relatives. This implies that those genes were also active in a common ancestor, and were killed off in some descendants but not in others. Out of about thirty thousand genes, for example, we humans carry more than two thousand pseudogenes. Our genome—and that of other species—are truly well populated graveyards of dead genes."[29]

Richard Dawkins continued to rely on junk DNA in his 2009 book *The Greatest Show on Earth: The Evidence for Evolution*. "It is a remarkable fact," Dawkins wrote, "that the greater part (95 per cent in the case of humans) of the genome might as well not be there, for all the difference it makes." In particular, pseudogenes "are genes that once did something useful but have now been sidelined and are never transcribed or translated." Dawkins concluded: "What pseudogenes are useful for is embarrassing creationists. It stretches even their creative ingenuity to make up a convincing reason why an intelligent designer should have created a pseudogene… unless he was deliberately setting out to fool us."[30]

In 2010, University of California Distinguished Professor of Ecology & Evolutionary Biology John C. Avise published a book titled *Inside the Human Genome: A Case for Non-Intelligent Design*, in which he wrote that "noncoding repetitive sequences—'junk DNA'—comprise the vast bulk (at least 50%, and probably much more) of the human genome." Avise argued that pseudogenes, in particular, are evidence against intelligent design. For example, "pseudogenes hardly seem like genomic features that would be designed by a wise engineer. Most of them lie scattered along the chromosomes like useless molecular cadavers." To be sure, "several instances are known or suspected in which a pseudogene formerly assumed to be genomic 'junk' was later deemed to have a functional role in cells. But such cases are almost certainly exceptions rather than the rule. And in any event, such examples hardly provide solid evidence for intelligent design; instead, they seem to point toward the kind of idiosyncratic tinkering for which nonsentient evolutionary processes are notorious."[31]

Avise also published an article in *Proceedings of the National Academy of Sciences USA* titled "Footprints of nonsentient design inside the human genome," in which he repeated the same argument. "Several outlandish features of the human genome," he wrote, "defy notions of ID by a caring cognitive agent," but they are "consistent with the notion of nonsentient contrivance by evolutionary forces." For example, "the vast majority of human DNA exists not as functional gene regions of any sort but, instead, consists of various classes of repetitive DNA sequences, including the decomposing corpses of deceased structural genes."[32]

But Is It True?

THE ARGUMENTS by Dawkins, Miller, Shermer, Collins, Kitcher, Coyne and Avise rest on the premise that most non-protein-coding DNA is junk, without any significant biological function. Yet a virtual flood of recent evidence shows that they are mistaken: Much of the DNA they claim to be "junk" actually performs important functions in living cells.

The following chapters cite hundreds of scientific articles (many of them freely accessible on the Internet) that testify to those functions—and those articles are only a small sample of a large and growing body of literature on the subject. This does not mean that the authors of those articles are critics of evolution or supporters of intelligent design. Indeed, most of them interpret the evidence within an evolutionary framework. But many of them explicitly point out that the evidence refutes the myth of junk DNA.

3.

Most DNA Is Transcribed into RNA

WHEN FRANCIS CRICK PROPOSED IN 1958 THAT DNA CONTROLS protein synthesis through the intermediary of RNA, he argued that "the transfer of information from nucleic acid to nucleic acid, or from nucleic acid to protein may be possible, but transfer from protein to protein, or from protein to nucleic acid, is impossible." Under some circumstances RNA might transfer sequence information to DNA, but the order of causation is normally "DNA makes RNA makes protein." Crick called this the "Central Dogma" of molecular biology.[1]

If DNA makes RNA makes protein, and one assumes that only protein-coding regions of DNA matter to the organism, it makes sense also to assume that only protein-coding regions are transcribed into RNA. Why would an organism struggling to survive waste precious internal resources on transcribing "junk"? Yet it turns out that organisms *do* transcribe most of their DNA into RNA—including DNA long regarded as junk. As we shall see, this calls into question arguments based on so-called "junk DNA."

DNA Makes RNA Makes Protein

THE GENERAL mechanism by which DNA makes RNA makes protein is now well understood. An enzyme called RNA polymerase moves along the DNA, transcribing the sequence of nucleotide subunits into messenger RNA—a process called "transcription." A large molecular machine called a ribosome then moves along the messenger RNA and translates it into a protein—a process called "translation." The process by which a DNA sequence yields a functional product (in this case a protein) is called "gene expression." (**Figure 3.1**)

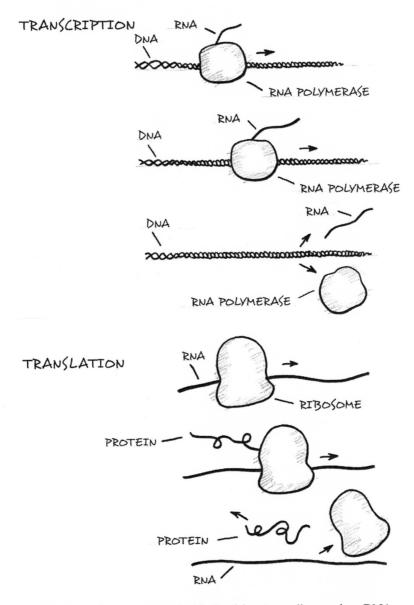

Figure 3.1 *Gene expression.* An idealized drawing to illustrate how DNA makes RNA makes protein. *Transcription.* (On the top) RNA polymerase moves along the DNA from left to right, producing a messenger RNA transcript (single line curving upward and to the right). *Translation.* (On the bottom) The bell-shaped ribosome moves along the messenger RNA transcript from left to right, translating it into a protein (curly line to the left).

As we saw in Chapter 2, many biologists in the 1970s equated Darwinian genes with DNA sequences. An organism's genes constitute its "genotype," while its morphology, physiology, development and behavior constitute its "phenotype." Each gene consists of a "promoter" section to which the RNA polymerase attaches, an "initiation sequence" and a "termination sequence." The actual protein-coding region is called an "open reading frame." **(Figure 3.2)**

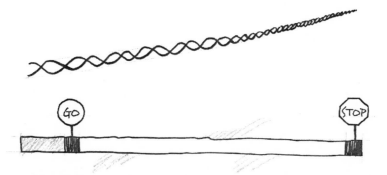

Figure 3.2 *Structure of an idealized gene.* The light gray block at the left is the "promoter," a sequence that responds to signals that turn the gene on or off. The black block on the left is the "initiation" sequence to which RNA polymerase attaches to begin making an RNA transcript (see Figure 3.1). The black block on the right is the "termination" sequence that releases the RNA polymerase and ends transcription. The long stretch between the initiation and termination sequences is the "open reading frame"—the DNA sequence that encodes the RNA sequence in the transcript.

Non-Protein-Coding DNA

IN THE mid-1970s, Richard Roberts and Phillip Sharp (studying viruses that cause respiratory infections) and David Glover and David Hogness (studying fruit flies) found evidence that open reading frames in eukaryotic genes are discontinuous: Protein-coding segments are separated by non-protein-coding segments.[2–4] (A eukaryote is a cell with a nucleus, as in animals and plants; a prokaryote is a cell without a nucleus, as in bacteria.) In 1978, Walter Gilbert called the protein-coding regions "exons" (**EX**pressed regi**ONS**) and the non-protein-coding regions "introns"

(**INTR**agenic regi**ONS**).[5] It soon became clear that most eukaryotic genes contain introns. (**Figure 3.3**)

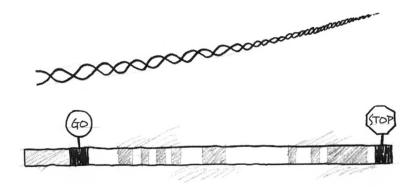

Figure 3.3 *Structure of an idealized eukaryotic gene.* The promoter, initiation site and termination site are similar to those in Figure 3.2, but the open reading frame is broken up into "exons" (white areas) and "introns" (gray areas between exons). The entire open reading frame is transcribed into RNA, but the RNA segments transcribed from introns are edited out; they are not translated into protein. Only the RNA segments corresponding to exons are translated into protein.

Non-protein-coding DNA in eukaryotes occurs not only within genes, but also between them. The two complementary strands of DNA can be separated in a test tube. Under appropriate conditions, the two strands will re-associate, though it takes some time for the complementary nucleotides on the two strands to align properly. In the 1960s, Roy Britten and others found that about 10% of mouse DNA re-associated extremely rapidly. When the researchers centrifuged the DNA to separate it into fractions of different densities, the fraction that re-associated rapidly ended up in "satellite" bands. These bands were found to consist of millions of short, repeated nucleotide sequences that do not code for proteins. Subsequent experiments showed that non-protein-coding repetitive sequences are common in animal DNA.[6–8]

In fact, only about 1.5% of human DNA codes for protein.[9] Eukaryotic chromosomes contain vast stretches of non-protein-coding DNA. (**Figure 3.4**) It was this preponderance of non-protein-coding regions that fueled the notion of "junk DNA" in the 1970s. The preponderance of non-protein-coding DNA also meant that the classical notion of genotype did not encompass all of an organism's DNA, and the word "genome" (which originally meant the same as genotype) was expanded to mean the complete DNA of an organism, including its non-protein-coding portions.[10]

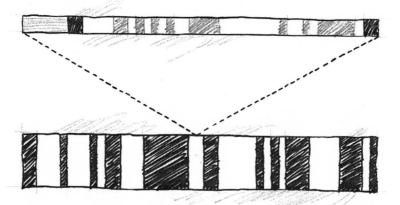

Figure 3.4 *Regions of protein-coding and non-protein-coding DNA.* (Top) The eukaryotic gene shown in Figure 3.3, with protein-coding exons (white) separated by non-coding introns (dark gray). (Bottom) A portion of an idealized eukaryotic chromosome, showing bands called euchromatin (white) that have a high concentration of protein-coding genes and bands called heterochromatin (black) that have a low concentration of genes. Even the euchromatin contains long stretches of non-protein-coding DNA between genes. The dotted lines indicate the position of the idealized gene in one band of euchromatin.

Genome Sequencing Projects

THE HUMAN Genome Project started in 1990 with the goal of cataloging the entire sequence of nucleotides (a little over three billion of them) in our DNA.[11] Sequences from humans and many other organisms are now catalogued at GenBank, a division of the National Center for Biotechnology Information in the United States;[12] at the European Molecular Biology Laboratory (EMBL) Nucleotide Sequence Database in the United Kingdom;[13] and at the DNA Data Bank of Japan (DDBJ).[14]

The Human Genome Project was completed in 2003, but the mere catalog of nucleotide sequences failed to explain how our DNA functions.[15] Looking at the sequence of the human genome is a bit like holding up a strip of videotape with its magnetic domains made visible: Knowing the coded information doesn't enable us to watch the movie. So a second project (called ENCODE, for ENCyclopedia Of DNA Elements) set out to identify all the functional elements in the human genome.[16] A similar project was undertaken by the FANTOM (Functional ANnotation Of the Mammalian Genome) Consortium of the Riken Institute in Japan, which had been founded in 1998.[17-18] By cataloging the functional products of the genome, both projects hoped to bring us closer to being able to watch the "movie" it encodes.

Even before completion of the Human Genome Project there had been reports of widespread transcription of RNA from non-protein-coding DNA. Despite the assumption that only protein-coding DNA matters to the organism and thus would be transcribed, American scientists estimated in 2001 that human DNA produces over 65,000 RNAs, with only about 4% of these coming from exons.[19] In 2002, the FANTOM Consortium identified 11,665 non-protein-coding RNAs and concluded that "non-coding RNA is a major component of the transcriptome."[20] (An organism's transcriptome is the entirety of its RNA.) Other scientists reported that transcription of two human chromosomes resulted in ten times more RNA than could be attributed to protein-coding exons.[21]

A few years after the start of the ENCODE Project it had become obvious that most of the mammalian genome is transcribed into RNA.[22-23] Preliminary data provided "convincing evidence that the genome is pervasively transcribed, such that the majority of its bases can be found in primary transcripts, including non-protein-coding transcripts."[24]

Even more surprising than the sheer number of transcripts was the complexity of the transcriptome. Molecular biologists originally thought that only one strand of the double-stranded DNA molecule (called the "sense" strand) carries information that is transcribed into

RNA. The other ("antisense") strand was thought to function only in DNA replication: The two strands separate before cell division, a new antisense strand is synthesized with a sequence complementary to that on the sense strand, and a new sense strand is synthesized with a sequence complementary to that on the antisense strand. **(Figure 3.5)**

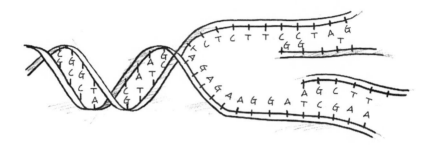

Figure 3.5 *DNA replication.* (Left) Double-stranded DNA. Because of their molecular structures, A's pair with T's and G's pair with C's. The sequence of nucleotides on one strand is thus complemented by the sequence of nucleotides on the other strand. (Middle) During replication the strands separate. (Right) New strands are synthesized by matching up complementary nucleotides. The result is two double-stranded DNA molecules with identical sequences (unless disrupted by mutations).

The ENCODE Project and FANTOM Consortium showed that RNAs are transcribed from *both* strands of DNA, and that "antisense" RNA is a major component of the transcriptome.[25–29]

Not only is RNA transcribed from the antisense strand, but RNAs can also be transcribed from multiple start sites within an open reading frame. As a result, a single open reading frame can generate, in addition to the primary protein-coding messenger RNA, several non-protein-coding RNAs.[30–33] **(Figure 3.6)**

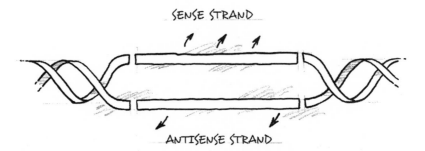

SENSE STRAND

ANTISENSE STRAND

Figure 3.6 *Sense and antisense transcription.* (Top) The sense strand of DNA. (Bottom) The antisense strand, previously thought to function only as a template for the replication of the sense strand. It is now known that both strands are transcribed into RNA, starting from multiple sites (arrows).

Probable Function in Non-Protein-Coding RNAs

WIDESPREAD TRANSCRIPTION of non-protein-coding DNA suggests that the RNAs produced from such DNA might serve biological functions. Ironically, the suggestion that much non-protein-coding DNA might be functional also comes from evolutionary theory. If two lineages diverge from a common ancestor that possesses regions of non-protein-coding DNA, and those regions are really nonfunctional, then they will accumulate random mutations that are not weeded out by natural selection. Many generations later, the sequences of the corresponding non-protein-coding regions in the two descendant lineages will probably be very different. On the other hand, if the original non-protein-coding DNA was functional, then natural selection will tend to weed out mutations affecting that function. Many generations later, the sequences of the corresponding non-protein-coding regions in the two descendant lineages will still be similar. (In evolutionary terminology, the sequences will be "conserved.") Turning the logic around, Darwinian theory implies that if evolutionarily divergent organisms share similar non-protein-coding DNA sequences, those sequences are probably functional.

In 2004 and 2005, several groups of scientists identified non-coding regions of DNA hundreds of nucleotides long that were *100% identical* in humans and mice. They called these "ultra-conserved regions (UCRs)" and noted that they clustered around genes involved in early development. The researchers concluded that the long non-coding UCRs act as regulators of developmentally important genes.[34–38]

In 2006, as part of a team studying endothelial cells (which line the inside of human blood vessels), Francis Collins co-authored a report that "conserved non-coding sequences"—some within introns—were enriched in sequences that "may play a key role in the regulation of endothelial gene expression."[39] In 2007, other scientists reported a clustering of highly conserved non-coding elements around developmentally important genes in worms and flies.[40]

Oxford geneticists comparing large non-coding RNAs in humans, rats and mice reported conserved sequences that "possess the imprint of purifying selection, thereby indicating their functionality."[41] And in 2009, a team of American scientists found "over a thousand highly conserved large non-coding RNAs in mammals" that are "implicated in diverse biological processes."[42]

Specific Functions in Non-Protein-Coding RNAs

EVEN APART from sequence conservation, there is growing evidence for specific functions of non-protein-coding RNAs. In 2003, Polish scientists reported that "non-protein-coding RNAs are known to play significant roles," primarily involving the regulation of gene expression. For example, non-protein-coding RNAs are involved in "controlling whether a gene is transcribed and to what extent," or "regulating the fate of the transcribed RNA molecules."[43]

In 2006, Australian molecular biologists noted that although exploring the functions of non-protein-coding RNAs had just begun, "these RNAs (including those derived from introns) appear to comprise a hidden layer of internal signals that control various levels of gene expression in physiology and development." Indeed, they wrote, "RNA regulatory networks may determine most of our complex characteristics."[44] Spanish scientists reported that small non-protein-coding RNAs "regulate virtually all aspects of the gene expression pathway, with profound biological consequences."[45]

In 2007, a team of American and Israeli scientists published evidence that developmental genes in humans produce, in addition to proteins, non-protein-coding RNAs that are spatially expressed in a developing embryo. The results, they wrote, "have broad implications for gene regulation in development."[46] By 2008, the scientific literature contained abundant data regarding the functions of non-protein-coding RNA.[47–51] One group of molecular biologists in Japan noted that since "research in the recent few years has identified an unexpectedly rich variety of mechanisms by which non-coding RNAs act," it is likely "that we have identified probably only a few of the many potential functional mechanisms" of the mammalian transcriptome.[52]

One recently identified function for non-protein-coding RNAs involves domains inside the nuclei of mammalian cells called "paraspeckles."[53] Paraspeckles play a role in gene expression by retaining certain RNAs within the nucleus,[54–55] and several non-protein-coding RNAs are known to be essential constituents of them.[56–57] The RNAs serve a structural function, binding to specific proteins to form ribonucleoproteins that stabilize the paraspeckles and enable them to persist through cell division even though they are not bounded by membranes.[58–59]

Evidence for important biological functions of non-protein-coding RNAs has continued to accumulate.[60–62] As the next chapter demonstrates, this includes evidence from introns, the non-protein-coding segments that separate protein-coding exons in a gene.

4.

INTRONS AND THE SPLICING CODE

WHEN A EUKARYOTIC GENE IS TRANSCRIBED INTO RNA, ITS INtrons as well as its exons are included in the transcript, so the initial RNA transcript consists of protein-coding segments separated by non-protein-coding segments. The latter are removed, and the protein-coding segments are then spliced together before being translated into protein. In the great majority of cases (80–95%), the protein-coding segments can be "alternatively spliced," which means that the resulting transcripts can lack some exons or contain duplicates of others.[1-7] (**Figure 4.1**) In this way, a single gene can give rise to hundreds—or even thousands—of different proteins.[8-10]

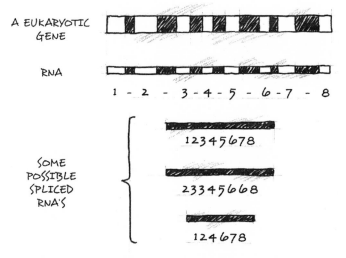

Figure 4.1 *Alternative RNA splicing.* (Top line) A eukaryotic gene. (Second line) The RNA transcribed from it. The transcript consists of protein-coding exons (numbers) separated by non-protein coding introns (dashes). (Third line) The RNA produced if the introns are simply removed. (Fourth and fifth lines) Exons can be duplicated or deleted to produce these or other RNAs. In this way a single gene can give rise to hundreds of different proteins.

Alternative Splicing Produces Tissue- and Stage-Specific RNAs and Proteins

ALTERNATIVE SPLICING plays an essential role in the differentiation of cells and tissues at the proper times during embryo development.[11–13] For example, in 2007 a British medical researcher reported that genes involved in triggering labor contractions are "both temporally and spatially regulated" by alternative splicing.[14] Two other British researchers reported that a crucial cell-cell signaling mechanism in animal embryos is mediated by alternative splicing in a "tissue- and stage-specific" manner.[15]

In 2009, Italian biologists found that a mammalian insulin-receptor gene is alternatively spliced into two proteins; one is predominantly active in fetuses and the other one in adults.[16] The same year, a team that included Francis Collins studied alternative splicing in various types of human cells (pancreas, colon, liver, blood, muscle, and fat) and reported that splicing is tissue-specific.[17]

In 2010, medical researchers published evidence that alternative splicing plays an essential role in brain development by producing variant forms of neurotransmitters[18] and proteins involved in intracellular transport.[19] German scientists showed that alternatively spliced forms of a gene involved in mouse mammary gland development were expressed in different tissues,[20] and Australian biologists reported that a wide variety of alternatively spliced RNAs occur in "a developmental-stage- and tissue-specific manner."[21] American and Canadian scientists found that alternative splicing is regulated, at least in part, by non-protein-coding RNAs.[22]

But what about introns? They make alternative splicing possible, but are they just biologically inert spacers? Apparently not; there is growing evidence that introns perform various functions—including the regulation of alternative splicing.

Evidence That Introns Help to Regulate Alternative Splicing

As WE saw in Chapter 3, evolutionary theory suggests that regions of non-protein-coding DNA that are similar between distant species were probably "conserved" by natural selection because they have some function; otherwise, mutations would have accumulated in the course of evolution and made them very different. In 2003, Israeli scientists compared alternatively spliced exons in humans and mice and found that over three-quarters of them were flanked by introns with sequences that were 80–88% conserved—suggesting that the introns function in the regulation of alternative splicing.[23]

In 2005, biologists at Lawrence Berkeley National Laboratory reported that a particular sequence of six nucleotides in introns that is "frequently located adjacent to tissue-specific alternative exons in the human genome" is "highly conserved" in species as distantly related as humans, mice, rats, dogs and chickens. The Berkeley scientists concluded that the sequence specificity, genomic location, and evolutionary conservation of this intronic element "mark it as a critical component of splicing switch mechanism(s) designed to activate a limited repertoire of splicing events in cell type-specific patterns."[24] In 2006, another group of California scientists identified intron sequences in brain and muscle tissues that were highly conserved among mammals, implicating them in splicing regulation.[25]

Sequence conservation suggests function in general, but there is also specific evidence that introns contain codes that regulate alternative splicing.[26-28] The mammalian thyroid hormone receptor gene produces two variant proteins with opposite effects, and the alternative splicing of those variants is regulated by an intron.[29] An intronic element plays a critical role in the alternative splicing of tissue-specific RNAs in mice,[30] and regulatory elements in introns control the alternative splicing of growth factor receptors in mammalian cells.[31]

In 2007, Italian biologists reported that intronic sequences regulate the alternative splicing of a gene involved in human blood clotting.[32] In 2008, American scientists summarized some of the splicing regulatory elements known to be located in introns,[33] and Scottish and French scientists reviewed intronic non-protein-coding RNAs that are involved in alternative splicing in plants as well as animals.[34]

In 2010, two American researchers identified splicing regulatory elements from the same intron that can have opposite effects in different tissues,[35] and another two reported "genome-wide evidence for exons being defined through the combinatorial activity of motifs located in flanking intronic regions."[36] A team of Canadian and British scientists studying splicing codes in mouse embryonic and adult tissues—including the central nervous system, muscles, and the digestive system—found that introns are rich in splicing-factor recognition sites. It had previously been assumed that most such sites are close to the affected exons—leaving long stretches of DNA not involved in the process of alternative splicing—but the team concluded that their results suggested "regulatory elements that are deeper into introns than previously appreciated."[37]

Other Coding Functions of Introns

INTRONS ARE also involved in gene regulation in ways other than alternative splicing. In 2007, European biologists found eleven sequences in the introns of a gene involved in organ development that were conserved from pufferfish to humans. Those sequences were part of larger conserved non-protein-coding elements that—when put into cultured human cells—acted as "cell type-specific enhancers of gene transcription."[38] In 2008, Brazilian researchers compared non-protein-coding RNAs from introns in humans and mice. The researchers found that not only the sequences but also the tissue-specific expression patterns were evolutionarily conserved; they concluded that such RNAs were "likely to be involved in the fine tuning of gene expression regulation in different mammalian tissues."[39] And a multinational group of scientists reported in 2009 that numerous small non-protein-coding RNAs involved in gene regulation in mammals and chickens showed "evolutionarily stable associations" with their host genes that suggested a role in regulating the expression of those genes.[40]

Short non-protein-coding RNAs are known to regulate gene expression,[41] and in 2004 British scientists identified such RNAs within the introns of 90 protein-coding genes.[42] In 2005, M.I.T. scientists described short RNAs that originate within the introns of the genes whose splicing they regulate.[43] In 2007, Korean biologists reported that in humans a "majority" of short non-protein-coding RNAs originate "within intronic regions."[44] One of these, according to American medical researchers, is involved in regulating cholesterol levels.[45]

As we saw in Chapter 3, messenger RNAs are translated into proteins by complex molecular machines called "ribosomes," which themselves are made up of proteins and long RNAs. Introns encode many of the small RNAs essential for the processing of ribosomal RNAs, as well as the regulatory elements associated with such RNA-coding sequences.[40,46]

Enhancers are DNA sequences involved in gene regulation that may be tens of thousands of nucleotides away from the genes they regulate.[47] In 2007, biologists determined that an enhancer of a gene involved in development in fishes and humans is encoded in sequences distributed throughout the gene's introns.[38] The following year, researchers studying a human gene involved in cartilage production likewise discovered an enhancer in one of the gene's introns.[48] In 2009, biologists reported finding an enhancer in an intron of a gene involved in chloride transport,[49] and in 2010 an enhancer was identified in an intron of a gene involved in milk production.[50]

Chromatin—the material of chromosomes—consists of a complex combination of DNA, RNA and proteins. If the DNA in a human cell were straightened out it would be about 3 meters long. To be contained within a cell, the DNA must be compacted in chromatin, and the first level of compaction involves winding the DNA around small spools made of proteins called "histones." These are then stacked together to produce the three-dimensional structure of the chromosome itself. (**Figure 4.2**)

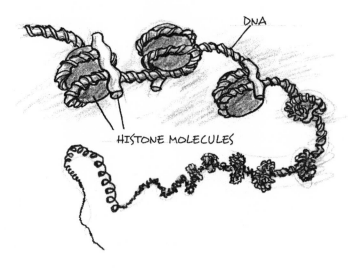

Figure 4.2 *Histones and chromatin.* The first level of structure in chromatin, with the long DNA molecule wrapped around small spools composed of proteins called histones.

Chromatin organization profoundly affects gene expression.[51–53] Non-protein-coding RNAs are essential for chromatin organization,[54–55] and non-protein-coding RNAs have been shown to affect gene expression by modifying chromatin structure.[56–57] Yet a recent study of chromatin-associated RNAs in some human cells revealed that almost two-thirds of them are derived from introns.[58]

The timing of gene expression is crucial for a living organism, and introns contain codes that affect this timing. In 2007, biologists reported that in fruit flies the heat-sensitive splicing of an intron "is critical for temperature-induced adjustments in the timing of evening activity."[59] In 2009, Chinese scientists reported that the developmental timing of a set of cells in roundworms is regulated by an intronic element.[60]

Yet introns can be thousands of nucleotides long, and documented coding functions account for only a fraction of those nucleotides. Is the remaining DNA non-functional, or might it function in some other way?

Intron Length Might Affect Gene Expression

IN 1986, British biologist David Gubb suggested that the time needed to transcribe eukaryotic genes is a factor in regulating the quantity of protein they produce. He proposed that the sheer length of introns in some genes "would affect both the spatial and temporal pattern of expression of their gene products."[61] In 1992, American biologist Carl Thummel likewise argued that "the physical arrangement and lengths of transcription units can play an important role in controlling their timing of expression." For example, the very long introns in certain key developmental genes could delay their transcription, "consistent with the observation that they function later in development" than genes with shorter introns.[62]

In 2008, Harvard systems biologists Ian Swinburne and Pamela Silver summarized circumstantial empirical evidence that intron length has significant effects on the timing of transcription. "Developmentally regulated gene networks," they wrote, "where timing and dynamic patterns of expression are critical, may be particularly sensitive to intron delays."[63]

So introns might have a function in gene regulation that is independent of exact nucleotide sequence. Although this remains to be demonstrated directly, there is already evidence that non-protein-coding DNA might also function in other ways that are independent of the precise order of its subunits. Chapter 7 will survey some of that evidence. First, however, we turn to a form of so-called "junk DNA" known as pseudogenes.

5.

PSEUDOGENES—NOT SO PSEUDO AFTER ALL

IN THE 1970S, MOLECULAR BIOLOGISTS FOUND A REGION OF DNA IN frogs that contained apparently inactive copies of a sequence that elsewhere (or in other organisms) coded for protein. They called the non-protein-coding copies "pseudogenes,"[1] and thousands of other pseudogenes have since been found in humans and other eukaryotes.[2–4] Indeed, the mammalian genomes studied so far have almost as many pseudogenes as they have protein-coding genes.[5]

As we saw in Chapter 2, pseudogenes are popular with writers trying to prove that Darwinian evolution is true and intelligent design is false. Kenneth Miller called them "discarded sequences" that are "consistent with an evolutionary explanation but inconsistent with intelligent design."[6] Douglas Futuyma wrote that only Darwinian evolution "can explain why the genome is full of 'fossil' genes: pseudogenes that have lost their function"—a phenomenon that he argues is "hard to reconcile with beneficent intelligent design."[7] According to Jerry Coyne, "the evolutionary prediction that we'll find pseudogenes has been fulfilled—amply," since "our genome—and that of other species—are truly well populated graveyards of dead genes."[8]

Richard Dawkins called pseudogenes "genes that once did something useful but have now been sidelined and are never transcribed or translated," and he concluded: "What pseudogenes are useful for is embarrassing creationists. It stretches even their creative ingenuity to make up a convincing reason why an intelligent designer should have created a pseudogene… unless he was deliberately setting out to fool us."[9] And John Avise wrote, "pseudogenes hardly seem like genomic features that would be designed by a wise engineer. Most of them lie scattered along the chromosomes like useless molecular cadavers" and "point toward the kind of idiosyncratic tinkering for which nonsentient evolutionary processes are notorious."[10]

Yet there is growing evidence that many pseudogenes are not functionless, after all.

Types of Pseudogenes

PSEUDOGENES ARE divided into three categories. (1) *Disabled (or unitary) pseudogenes* are single sequences that may have once coded for protein but have apparently been inactivated by nucleotide changes or deletions. (2) *Duplicated pseudogenes* are copies of still-functioning genes, though unlike the functioning originals they have characteristics that prevent them from encoding proteins. (3) *Processed pseudogenes* have sequences similar to those of functioning genes, except that they lack promoter sequences and are usually missing introns.[11] **(Figure 5.1)**

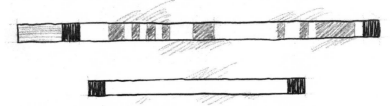

Figure 5.1 *A processed pseudogene.* (Top) Structure of an idealized eukaryotic gene like that shown in Figure 3.3, with a promoter (light gray box at far left), initiation and termination sites (black boxes), exons (white boxes) and introns (gray boxes separating exons). (Bottom) An idealized processed pseudogene, with a protein-coding sequence similar to the one in the gene above but lacking a promoter and introns.

Since introns are edited out of messenger RNA sequences before the latter are translated into proteins, the absence of introns in processed pseudogenes suggests that they were "reverse transcribed" from messenger RNA back into DNA—a process called "retrotransposition."[12] (More about this in Chapter 6.) The majority of pseudogenes fall into this third category, processed pseudogenes.

Transcribed Pseudogenes

EVIDENCE THAT many pseudogenes are transcribed into RNA began accumulating in the 1990s. Specific examples in humans include pseudogenes corresponding to genes involved in carbohydrate and lipid metabolism,[13–14] a gene involved in regulating estrogen levels,[15] a gene involved in the process of protein synthesis,[16–17] and a gene involved in muscle movements.[18] Examples in cows include pseudogenes that correspond to a gene involved in basic metabolism[19] and a gene involved in estrogen synthesis.[20] Examples in plants include pseudogenes corresponding to protein components of ribosomes, the molecular machines that translate RNAs into proteins.[21–22]

Since 2000, evidence for pseudogene transcription has been accumulating rapidly. In one study, biologists working with the ENCODE Project sampled 201 pseudogenes and found that at least one-fifth of them are transcribed in one or more tissues.[23–25]

Some pseudogene-encoded RNAs have characteristics suggesting that they may be capable of being translated into protein. Examples in humans include pseudogenes corresponding to genes for a molecule involved in the immune system,[26] a neurotransmitter,[27] a neurotransmitter receptor,[28] a DNA-binding protein,[29] and a membrane protein involved in cell-cell communication.[30]

In fact, it is now known that a few pseudogene-derived RNAs actually *are* translated into proteins.

"Pseudogenes" That Encode Proteins

IN 1988, Swiss scientists found a human gene lacking introns and concluded that it was a pseudogene.[31] A few years later, however, American scientists discovered that the gene encodes a messenger RNA that is translated into protein.[32]

In 1991, British biologists studying an enzyme that detoxifies alcohol found intron-lacking genes for it in two species of fruit fly and concluded that they were processed pseudogenes.[33] Two years later, however,

American biologists reported that the putative pseudogenes produce a functional protein and thus are not pseudogenes after all.[34]

In 1996, biologists identified a gene in fruit flies that contained premature transcription termination sites, and they proposed that it "may be a pseudogene."[35] The following year, however, other biologists reported that it encodes a functional enzyme.[36]

In 1997, University of Michigan researchers identified a human gene "with the typical features of a processed pseudogene." When they were unable to find any expression of the gene they concluded that it was, indeed, a pseudogene.[37] In 2002, however, biologists at the University of Chicago and University of Cincinnati found evidence that the gene "is encoding a functional protein."[38]

In 2000, French biologists reported that a presumed pseudogene in cultured human melanoma cells actually produces functional protein.[39] It seems that in the case of pseudogenes (with apologies to Mark Twain), reports of their death have been greatly exaggerated.

To be sure, only a relatively small proportion of known pseudogenes have been shown to encode proteins. But there is growing evidence that RNAs transcribed from pseudogenes perform essential functions in the cell.

RNA Interference

IN THE 1990s, molecular biologists discovered that the antisense strands of some pseudogenes are transcribed into RNA, and they suggested that such RNA might play a role in regulating gene expression.[40–41]

Since the sequence of DNA in a processed pseudogene is very similar to the sequence of the protein-coding segments (exons) of the complete gene (**Figure 5.2**), its RNA mirrors a messenger RNA transcribed from the functional gene, minus its introns. So (in the absence of alternative splicing) the RNA transcribed from one strand of the pseudogene is complementary to the messenger RNA transcribed from the opposite strand of the functional gene.

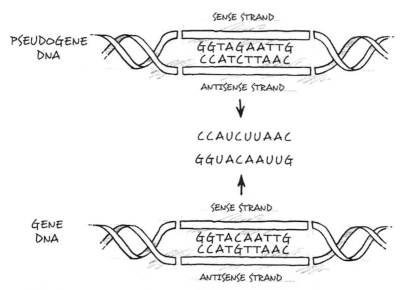

Figure 5.2 *RNA interference.* (Top) The double-stranded DNA of a pseudogene, showing the complementary nucleotide sequences in the sense and antisense strands. (Bottom) The equivalent portion of the corresponding gene. (Middle) RNAs transcribed from the pseudogene and the corresponding gene. (In RNA, U takes the place of T.) The two RNAs are not completely complementary (note that they both have a C in the fourth position from left), but they are close enough to being complementary that they bind to each other, forming a double-stranded RNA that interferes with the process of translation and reduces the amount of protein produced from the gene.

The two RNAs could bind together, much as the two complementary strands of DNA bind to each other. The result would be double-stranded RNA. But double-stranded RNA is not translated; instead, it interferes with translation and thereby reduces gene expression.[42–45] Cells make good use of RNA interference to regulate the amount of protein they produce.

In the 1990s, biologists in England found that the expression of a gene in the central nervous system of snails was "substantially suppressed" by antisense transcripts from a corresponding pseudogene. The pseudogene RNAs formed "duplex molecules" with the messenger RNAs from the gene itself, leading the biologists to suggest that transcribed pseudogenes "are a potential source of a new class of regulatory gene in the nervous system."[46]

A 2008 article in *Nature* reported that RNAs produced from pseudogenes regulate gene expression in mouse eggs by "RNA interference," in which double-stranded RNAs "suppress specific transcripts in a sequence-dependent manner."[47] The authors of an accompanying article concluded that their findings "indicate a function for pseudogenes in regulating gene expression by means of the RNA interference pathway."[48]

RNA that regulates gene expression can also be generated from a duplicated pseudogene (as opposed to a processed pseudogene). In 2009, biologists reported that small antisense RNAs derived from pseudogenes in rice were produced in specific developmental stages or physiological conditions, and they suggested that these "small interfering RNAs" probably had important roles in regulating gene expression.[49]

Pseudogene Enhancement of Gene Expression

PSEUDOGENE-ENCODED RNA may also *enhance* the expression of a protein-coding gene. In 2003, a team of Japanese and American biologists reported some experiments on the pseudogene corresponding to a mouse gene that encodes an enzyme called Makorin-1. They found that reducing the transcription of the pseudogene also reduced the expression of the gene itself, and they inferred that the pseudogene-derived RNA served to protect the *Makorin-1*-derived messenger RNA from degradation.[50–52]

Cells contain enzymes that degrade messenger RNAs to regulate the amount of protein transcribed from them. The longer a messenger RNA escapes degradation, the more protein molecules can be translated from it. The researchers in 2003 suggested that the pseudogene-derived RNA might provide an alternate target for the enzyme(s) that would normally degrade the *Makorin-1* messenger RNA, thus allowing continued translation of the latter.

Another possibility, suggested at the time by Harvard geneticist Jeannie Lee, was that the pseudogene-derived RNA functioned by blocking a repressor of the *Makorin-1* gene.[53] (Other biologists later challenged the *Makorin-1* pseudogene results, which remain controversial.)[54–55]

In 2007, European biologists reported that the expression of a plant pseudogene increased the expression of a protein-coding gene involved in phosphorus metabolism. They found that the pseudogene produced an RNA that provided an alternative target for a molecule that would normally have repressed translation of the messenger RNA from the protein-coding gene, and they coined the term "target mimicry" to describe the process.[56]

In 2008, a team of Norwegian and German biologists suppressed transcription of the pseudogene corresponding to a gene involved in transporting molecules across membranes, and they found that the expression of the functional gene was reduced as well. In other words, normal expression of the protein-coding gene depended somehow on transcription of the pseudogene. The team concluded that this provided evidence "for a regulatory interdependence of a transcribed pseudogene and its protein coding counterpart in the human genome," though they did not know the exact mechanism.[57]

In 2010, American biologists reported that the expression of two human genes is increased by transcription of their related pseudogenes. They traced the effect to pseudogene-derived RNA transcripts that serve as "perfect decoys" for molecules that would otherwise repress the protein-coding genes, and they concluded that "pseudogenes have an intrinsic biological activity" in regulating gene expression.[58]

The Vitamin C Pseudogene

ONE PARTICULAR pseudogene plays a prominent role in the arguments of Kenneth Miller and Jerry Coyne: the vitamin C pseudogene. Vitamin C is essential for many biochemical reactions in living cells, and its synthesis requires four enzymes. The human genome has only three of these; it also contains a segment of DNA very similar to the gene for the fourth enzyme, but this segment of DNA is not translated into protein.[59–60] In other words, the human genome contains a vitamin C pseudogene.

As we saw in Chapter 2, Miller and Coyne both argue that the vitamin C pseudogene provides evidence for Darwinian evolution—in particular, for the common ancestry of humans and other primates—and evidence against intelligent design or creation. The evidence is not as straightforward as Miller and Coyne make it out to be, however, and their argument is ultimately circular. In any case, common ancestry and intelligent design are two different issues, and the vitamin C story would take us on a detour from the issue of junk DNA that is the focus of this book, so the details are omitted here and included in an appendix.

Sequence Conservation

As we saw in Chapter 3, Darwinian theory predicts that nonfunctional DNA will accumulate damaging mutations over time. Thus similar ("conserved") sequences in the non-protein-coding DNA of evolutionarily distant organisms imply that such DNA is functional. This same logic has been applied to pseudogenes.[61]

In 2003, Evgeniy Balakirev and Francisco Ayala reviewed sequence data from humans, mice, chickens and fruit flies and reported "pseudogene features that would be unexpected if pseudogenes were nonfunctional sequences of genome DNA ('junk' DNA)." In particular, they found that "pseudogenes are often extremely conserved," implying that they are subject to natural selection and not free to accumulate random mutations. Balakirev and Ayala regarded this (along with widespread transcription) as evidence that many pseudogenes are not functionless, after all.[62]

In 2009, Canadian biologists Amit Khachane and Paul Harrison compared pseudogenes in humans, monkeys, mice, rats, dogs and cows and found significant sequence similarity, implying that the pseudogenes had been conserved by natural selection. They concluded that "through evolutionary analysis, we have identified candidate sequences for functional human transcribed pseudogenes."[63]

How odd! As we saw in Chapter 2, Kenneth Miller, Richard Dawkins, Douglas Futuyma, Michael Shermer, Jerry Coyne and John Avise argue that pseudogenes confirm Darwinism because they are non-functional. But if we assume that Darwinism is true and then compare the DNA of unrelated organisms, sequence similarities imply that many of their pseudogenes are functional. So nonfunction supposedly implies Darwinism, but Darwinism plus sequence conservation implies function. When it comes to conserved pseudogenes, it seems, Darwinism saws off the very branch on which it sits.

In the next chapter, we turn to one of the most commonly cited sources of evidence for so-called "junk DNA"—repetitive DNA.

6.

JUMPING GENES AND
REPETITIVE DNA

A LARGE PROPORTION OF NON-PROTEIN-CODING DNA CONSISTS OF movable and repetitive sequences. As we saw in Chapter 2, Kenneth Miller wrote that "the human genome is littered" with "so many repeated copies of pointless DNA sequences that it cannot be attributed to anything that resembles intelligent design." According to Richard Dawkins, much of DNA "consists of multiple copies of junk, 'tandem repeats,' and other nonsense," which "doesn't seem to be used in the body itself." Francis Collins acknowledged in 2006 that some repetitive elements may be functional, but he argued that most have no function other than their own survival and thus provide compelling support for Darwinian evolution. John Avise wrote that "several outlandish features of the human genome defy notions of ID by a caring cognitive agent," but they are "consistent with the notion of nonsentient contrivance by evolutionary forces." For example, "the vast majority of human DNA exists not as functional gene regions of any sort but, instead, consists of various classes of repetitive DNA sequences." Yet there is growing evidence that a great deal of repetitive DNA is transcribed into functional RNAs.

Jumping Genes

EVEN BEFORE Watson and Crick discovered the structure of DNA in 1953, Barbara McClintock had discovered "jumping genes" in corn. The varied colors of the kernels in a single ear of maize, she found, are due to mobile genetic elements called "transposons," which move from one place in the genome to another.[1-3] Some of these are segments of DNA that have been moved by a "cut and paste" process called "transposition." (Figure 6.1)

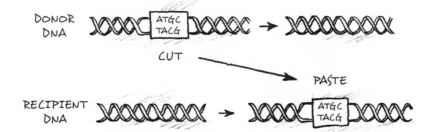

Figure 6.1 *Cut-and-paste transposition.* (Top left) Double-stranded DNA containing a simplified transposon (box). Actual transposons are much longer. (Top right) After the transposon is cut out, the DNA is shorter. (Bottom left) The segment of DNA into which the transposon will be pasted, which may be on the same DNA molecule or a different one. (Bottom right) The recipient DNA after the transposon has been pasted into it.

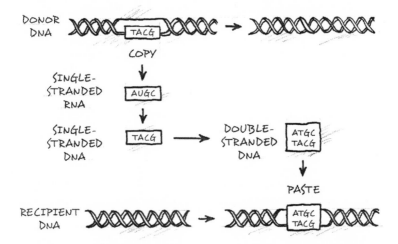

Figure 6.2 *Copy-and-paste retrotransposition.* (Top left) Double-stranded DNA containing a simplified retrotransposon (box). Actual retrotransposons are much longer. (Middle left) Single-stranded RNA copied (transcribed) from the retrotransposon (with U replacing T). This RNA is then reverse-transcribed into single-stranded DNA. (Middle right) The nucleotides in the single-stranded DNA pair with complementary nucleotides to make double-stranded DNA. (Bottom right) The new double-stranded DNA is then pasted into the recipient, which may be on the same DNA molecule or a different one. The recipient DNA is longer after pasting, but the length of the donor DNA (Top right) has not changed.

Other transposons (like processed pseudogenes) use RNA as an intermediary. In 1970, David Baltimore and Howard Temin independently discovered that RNA sequences can be transferred to DNA in a process called "reverse transcription."[4-5] After an enzyme nicks the organism's DNA, the newly "reverse transcribed" DNA is then inserted into that location in a "copy and paste" process called "retrotransposition." (**Figure 6.2**) Transposons that use RNA as an intermediate are called "retrotransposons," and they are a major component of repetitive DNA.

Types of Repetitive DNA

As we saw in Chapter 3, repetitive non-protein-coding DNA was discovered in the 1960s.[6-8] Repetitive DNA makes up about half of the human genome, and about two-thirds of repetitive DNA consists of retrotransposons that fall into two classes: Long Interspersed Nuclear Elements (LINEs) and Short Interspersed Nuclear Elements (SINEs).[9] (**Table I**)

TABLE I. SOME MAJOR COMPONENTS OF THE HUMAN GENOME

Approximate percentages of several types of DNA in the human genome.[10]

OPEN READING FRAMES ("GENES")	27%
Exons (Protein-coding regions)	2%
Introns (Non-protein-coding regions)	25%
REPETITIVE NON-PROTEIN-CODING DNA	50%
LINEs	21%
SINEs	13%
Retroviral-like elements	8%
Simple sequence repeats	5%
DNA-only transposons	3%
OTHER NON-PROTEIN-CODING DNA	23%

LINEs can be more than 5,000 nucleotides long, and some include DNA sequences encoding enzymes that enable them to reinsert themselves into DNA. Many LINEs also contain sense and antisense promoters. Mammalian genomes contain tens of thousands of LINEs, which fall into several groups; the most common is designated L1.

SINEs tend to be fewer than 500 nucleotides long and depend on other mobile genetic elements for their retrotransposition. The human genome contains over a million of them. The most common SINEs in primates are called *Alu* sequences because they are recognized by an enzyme from the bacterium *Arthrobacter luteus* (which is also why *Alu*, unlike the names of other repetitive DNA elements, is customarily italicized). *Alus* consist of about 300 nucleotides in a characteristic sequence.[11] The mouse genome contains SINEs with different sequences, designated B1, B2 and B4. The rat genome has a major SINE designated ID. It may be that every mammalian species has its own repertoire of SINEs.

Many LINEs and SINEs Are Functional

As WE saw in Chapter 3, most human DNA is transcribed; this includes repetitive DNA.[12–13] Such widespread transcription suggests that repetitive DNA might be functional. Indeed, plant molecular biologists reported in 2000 that "retrotransposons are central players in the structure, evolution and function of plant genomes"; they "are certainly not junk."[14]

As in the case of pseudogenes, the functionality of repetitive DNA has been inferred from evolutionary analyses. In 2006, scientists identified a family of SINEs that were "highly conserved" in mammals and concluded that they are functional.[15]

In 2007, biologists in California found that "the majority of conserved and, by extension, functional sequence in the human genome" seems to be outside of protein-coding exons and to consist of "mobile elements" of "clear repetitive origins."[16] Biologists in New York examined SINEs in humans and mice and reported in 2009 "that *Alu* and B1 elements have been selectively retained in the upstream [ahead of the promoter] and intronic regions of genes belonging to specific functional classes." Furthermore, "*Alu* and B1 elements show similar biases in their distribution across functional classes," strengthening the inference that they serve important biological functions.[17]

Widespread transcription and sequence conservation are not the only grounds for inferring the functionality of repetitive DNA. There is also a large and growing body of experimental evidence for specific functions of LINEs and SINEs, such as regulating the expression of other RNAs and the protein-coding regions of DNA.

Some Specific Functions of LINEs and SINEs

In mammals, males have a Y chromosome and an X chromosome, while females have two X chromosomes. In order for a female embryo to develop normally, one of the two X chromosomes must be inactivated.[18] In 2000, American biologists found evidence that X chromosomes are enriched in L1 LINE elements, and they suggested that LINEs are involved in the process of inactivation.[19] In 2010, British researchers reported that X chromosome inactivation depends on non-protein-coding RNAs that act more efficiently in L1-rich domains.[20] The same year, French biologists concluded that LINEs function at two different levels in X chromosome inactivation: First, LINE DNA produces a re-arrangement in the chromatin that inactivates some genes; second, RNAs transcribed from LINEs coat and silence other portions of the chromosome.[21]

In 2002, a team of American biologists reported that LINEs participate in repairing DNA breaks in cultured hamster cells.[22] Two members of that team, together with some other American scientists, reported in 2007 that human L1 sequences also function by mobilizing various RNAs in the cell.[23] The same year, British biologists showed that L1 elements are responsible for silencing a gene that is expressed in the liver in human fetuses but not in adults.[24]

In 2008, an Italian biologist reviewed the evidence and concluded that human L1 "regulates fundamental biological processes."[25] In 2009, Australian scientists reported that RNA transcribed from LINEs is an "essential structural and functional component" of "neocentromeres"[26]—

features of chromosomes that will be discussed in more detail in Chapter 7.

There is also abundant evidence for the functionality of SINEs. In a few cases, the protein-coding regions of active genes consist almost entirely of DNA sequences derived from mobile elements. Researchers found in 1985 that the protein-coding portion of one mouse gene is more than 90% similar to B2.[27] In 2004, Roy Britten reported that 99% of the coding sequence of one human gene expressed in brain cells consists of *Alu* sequences.[28]

In 1986, Russian scientists reported that B2 elements help to regulate the transcription of rat ribosomal RNA, an essential part of the cellular machinery that translates RNAs into proteins.[29] In 1999 and 2001, American scientists found that SINE RNAs in silkworms play "a role in the cell stress response" to heat or toxic chemicals.[30–31] Other researchers reported that B1 elements provide platforms for enzymes that regulate gene expression by chemically modifying (though not changing the sequence of) certain segments of DNA.[32] In 2004, American scientists showed that a B2 element in mice regulates transcription by blocking RNA polymerase.[33–34]

Alu elements contain functional binding sites for transcription factors.[35] RNAs derived from *Alu* sequences repress transcription during the cellular response to elevated temperatures.[36] *Alus* are also involved in the editing and alternative spicing of RNAs and in the translation of RNAs into proteins.[37–41] In 2009, Colorado researchers studying the biological functions of B2 and *Alu* SINEs reported that both types of repetitive DNA are transcribed into RNAs. The RNAs, in turn, help to control gene expression by controlling the transcription of messenger RNAs and by editing other RNAs. According to the researchers, "finding... that these SINE-encoded RNAs indeed have biological functions has refuted the historical notion that SINEs are merely 'junk DNA.'"[42]

SINEs can also influence transcription by affecting chromatin. When stained with appropriate chemicals and viewed under a light microscope, chromatin exhibits banding that is characteristic of a particular chromosome. The pattern resembles a bar code, like the lower part of Figure 3.4, and it includes two types of bands. One (called heterochromatin) is tightly packed and rich in the nucleotides A and T; it also has a low concentration of protein-coding sequences and a high density of L1 LINEs. The other (called euchromatin) is loosely packed and rich in the nucleotides G and C; it has a high concentration of protein-coding sequences and a high density of SINEs such as *Alus* or B1s and B2s.[43–45]

Swiss biologists who fed fruit flies a DNA-binding compound that targets repetitive sequences reported in 2000 that such sequences regulate gene expression by maintaining chromatin integrity.[46–47] American biologists studying fruit flies demonstrated that transposable elements are responsible for maintaining "telomeres"—the repetitive sequences at the ends of chromosomes that protect the latter from deterioration.[48–50]

In 2004, a team of French and American scientists studying a small flowering plant commonly called rock cress or thale cress (*Arabidopsis thaliana*) reported that its chromatin structure "is determined by transposable elements and related tandem repeats" that thereby contribute to gene regulation.[51] This regulation is due in part to RNA interference (Chapter 5).[52]

SINEs also help to regulate gene expression in mammalian development by establishing functional chromatin domains. In 2007, biologists reported that tissue-specific transcription of B2s is required for gene activation in developing mice. Their data suggested that "transcription of interspersed repetitive sequences may represent a developmental strategy for the establishment of functionally distinct domains within the mammalian genome to control gene activation."[53]

In 2010, biologists in India wrote that repetitive non-protein-coding DNA plays "a regulatory role by contributing to the packaging of the genome during cellular differentiation."[54] And Japanese biologists showed that untranscribed repeated copies of the DNA that codes for ribosomal RNA contribute to the cohesion of duplicated chromosomes before they separate during cell division.[55]

Argonaute, Piwi and RNA Silencing

IN THE 1990s, botanists found an *Arabidopsis* mutant that produces leaves resembling the tentacles of the small octopus *Argonauta argo*, and they named the mutant "argonaute." The effect was traced to a gene product that resembles proteins with unknown functions in animals ranging from worms to humans.[56]

Biologists soon discovered that the product of the gene affected by the argonaute mutation is involved in RNA interference. The argonaute protein is part of an "RNA-induced silencing complex" that regulates the expression of other genes by cutting up the messenger RNAs they produce.[57–62] Other components of the complex were given colorful names such as "Dicer" and "Slicer."[63–64]

In 1997, biologists used transposons called P elements to produce a mutation that abolished germline stem cell divisions in fruit flies, and they named the affected gene "*piwi*" (for "P element-induced wimpy testis").[65–66] Similar genes were found in worms, humans, and plants.[67] It turned out that the Piwi protein is part of the RNA-induced silencing complex.[68–70]

The Argonaute and Piwi proteins find their targets with the help of small non-protein-coding RNAs that are complementary to the target sequences. Many of those small RNAs are derived from repetitive DNA, including retrotransposons. This is true not only in fruit flies,[71–73] but also in mammals.[74–75]

In 2010, a team of French and American biologists reported that Piwi-associated RNAs and proteins act together to promote the timely decay of specific messenger RNAs in fruit fly embryos. Impairing this function of Piwi RNAs led to defects in head development. Because the Piwi RNAs "are produced from transposable elements," the team concluded, "this identifies a direct developmental function for transposable elements in the regulation of gene expression."[76]

Endogenous Retroviruses

MOST VIRUSES consist of DNA surrounded by a coat of protein that is encoded by that DNA. The virus infects a living cell by injecting its DNA into it; the cell's molecular machinery then makes copies of the viral DNA and synthesizes new protein coats; the replicated viruses are subsequently released to infect other cells. Some viruses, however, contain RNA instead of DNA. They inject their RNA into a living cell, and the cell then reverse transcribes the viral RNA into DNA. This virus-encoded DNA may be inserted into the cell's DNA, where it may then be transcribed into new viral RNA and new protein coats to make new viruses. Because RNA viruses are reverse transcribed inside the cell, they are called "retroviruses."[77–79]

In the early 1970s, biologists studying some chicken and quail cells that had not been infected with a particular retrovirus found that the cells nevertheless contained DNA sequences complementary to that virus's RNA.[80–81] Scientists assumed that the virus had infected the birds' ancestors, and that the viral DNA was then passed down from generation to generation as an "endogenous retrovirus" (ERV).[82]

DNA that is reverse transcribed from retroviral RNA is characteristically flanked by sequences that are repeated hundreds or thousands of times, called "long terminal repeats" (LTRs).[83–84] The LTR on one end of an ERV is in the same orientation as the LTR on the other end; thus endogenous retroviruses differ from DNA-only ("cut and paste") transposons, which are flanked by short inverted repeats.

At first glance, ERVs might seem to be a perfect example of "selfish DNA"— molecular parasites that hitch a ride in an organism's genome but perform no useful functions. It turns out, however, that many ERVs *do* perform useful functions. In the 1990s, French researchers reported that the transcription of a human gene involved in the production of blood cells[85] is regulated by the LTRs of an endogenous retrovirus.[86] A few years later, Canadian biologists reported that the LTRs of retroviral elements contain promoters that help to regulate the expression of human genes involved in fat metabolism and cell signaling in the liver and placenta.[87–88]

Subsequent research showed that ERVs contain promoters that regulate the expression of genes in mouse oocytes and early embryos[89–90] and in primate embryonic and blood-producing cells.[91] Human ERVs contain promoters that regulate genes involved in bicarbonate transport[92] as well as gene expression in the gastrointestinal tract, mammary glands, and testes.[93–95] Biologists from Asia, the U.S. and Europe have recently published additional evidence that promoters in the LTRs of human endogenous retroviruses contribute to cell-specific and tissue-specific gene expression.[96–98] The best-studied example is the placenta.

ERVs and Placentas

In the 1990s, British biologists studying the envelope protein of a human endogenous retrovirus discovered that it was both evolutionarily conserved and abundantly expressed in cells of the placenta. They concluded that the ERV has "a biological function."[99]

The placenta, which supplies nutrients to the embryo and serves as the interface between it and the mother, develops from "trophoblasts"— cells that are derived from the embryo and form a layer around it but are not incorporated into the fetus. In order for the placenta to function properly, some trophoblast cells must fuse into one giant, multinucleated cell, or "syncytium" (pronounced sin-SISH-um). (**Figure 6.3**)

In 2000, evidence suggested that the ERV envelope protein that is highly expressed in the placenta might be involved in the fusion of trophoblast cells, and the protein was named "syncytin" (pronounced sin-SIGHT-in).[100–101] Subsequent research confirmed the role of syncytin in the fusion of trophoblast cells during placental development.[102–104] Some women suffering from placental dysfunction were found to have reduced levels of syncytin.[105] On the other hand, people suffering from multiple sclerosis were found to have abnormally high expression of syncytin in cells that normally protect nerves.[106]

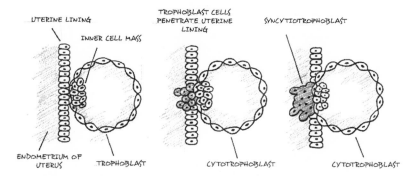

Figure 6.3 *Embryo implantation in mammals.* (Left) The early embryo contacts the inner wall of the uterus. (Middle) Outer cells from the embryo ("trophoblasts") migrate into the uterine lining. (Right) The trophoblast cells become a "syncytium," a single multi-nucleated cell that facilitates the transport of nutrients from the mother to the embryo.

In 2003, a team of French biologists reported finding a second ERV envelope protein involved in placenta development. They named it syncytin-2 and renamed the first syncytin-1.[107–108] French biologists also discovered two additional forms of the ERV envelope protein in mice and named them syncytin-A and syncytin-B.[109–110] And in 2009, French biologists discovered another form of syncytin in rabbits.[111] Surprisingly, although all the syncytins serve similar functions, syncytin-A and syncytin-B are unrelated to syncytin-1 and syncytin-2, and rabbit syncytin is unrelated to either the mouse or the human forms.

A British virologist in 2009 noted that it used to be "an open question" whether ERVs "simply represented junk or selfish DNA," but he called the work on syncytin-A and syncytin-B "compelling evidence" that at least some ERVs are making "a specific contribution to normal physiology."[112]

In addition to the part that encodes the protein, there are non-protein-coding parts of the syncytin ERV that are functional as well. In 2004, researchers determined that the long terminal repeat (LTR) of the ERV containing the syncytin gene contains the gene's promoter.[113–115]

Francis Collins and Repetitive Elements

As WE saw in Chapter 2, Francis Collins claimed in his 2006 book *The Language of God* that "ancient repetitive elements (AREs)" provide "compelling" evidence for Darwinian evolution, "with roughly 45 percent of the human genome made up of such flotsam and jetsam." The term "ancient repetitive element" is rarely used in the scientific literature, and Collins did not define precisely what he meant by it, but the "roughly 45 percent of the human genome" that he called repetitive "flotsam and jetsam" presumably included LINEs, SINEs, and ERVs—which, as we saw above, perform many biological functions.

Of course, there is much repetitive DNA for which functions have not yet been discovered, but when Collins published his book in 2006 there was already considerable evidence for the functionality of repetitive DNA. Indeed, a single review article published in 2005, titled "Why repetitive DNA is essential to genome function," described more than 80 known functions and cited over 200 scientific articles.[116]

Collins made particular use of repetitive elements as evidence for the common ancestry of humans and mice. "In many instances," he wrote, "one can identify a decapitated and utterly defunct ARE in parallel positions in the human and the mouse genome." These provide compelling support for Darwinian evolution, Collins argued, "unless one is willing to take the position that God has placed these decapitated AREs in these precise positions to confuse and mislead us."[117]

Collins's argument rests on the assumption that those repetitive elements (which he does not specifically identify) are nonfunctional. Yet their similar positions in the human and mouse genomes could mean that they are performing some function in both. Given the rate at which functions are being discovered, Collins's assumption seems foolhardy, and his argument could eventually collapse in the face of new scientific discoveries.

So far we have considered functions of so-called "junk DNA" that depend on the exact sequence of nucleotides in DNA or RNA. As we shall see in the next chapter, however, non-protein-coding DNA also functions in ways that are independent of its sequence.

7.

FUNCTIONS INDEPENDENT
OF EXACT SEQUENCE

A VERY SMALL PERCENTAGE OF OUR DNA FUNCTIONS BY ENCODING proteins; a much larger percentage functions by encoding RNAs with sequences that regulate gene expression and perform other roles in cells. But some of our DNA has functions that are independent of the exact sequence of nucleotide subunits.

Chapters 3–6 dealt mostly with sequence-dependent functions, though there were occasional hints of sequence-independent roles. For example, Chapter 4 cited biologists who think that introns regulate the timing of transcription, in part, simply by their length.[1-3] Chapter 6 listed some of the evidence that long and short repetitive elements (LINEs and SINEs) affect the large-scale organization of chromatin, which in turn affects gene expression.[4-7]

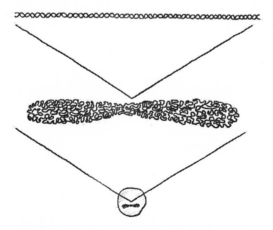

Figure 7.1 *The hierarchical structure of the genome.* (Top) The DNA molecule itself. (Middle) The chromatin (DNA, RNA, and protein) that makes up the chromosome. (Bottom) The position of the chromosome within the nucleus.

The genome functions in a hierarchical fashion. The DNA molecule is only the first level; chromatin organization is a second level; and the position of chromosomes within the nucleus is a third level.[8-9] (**Figure 7.1**) As we shall see, there is evidence at all three levels that non-protein-coding DNA performs functions that are independent of its exact sequence.

The First Level: The DNA Molecule

JUST AS introns might regulate the timing of transcription simply by their length, so the long stretches of non-protein-coding DNA between genes might affect their expression. In 1997, molecular biologist Emile Zuckerkandl emphasized that DNA may function in ways that do not depend on its particular nucleotide sequence. "Along noncoding sequences," he wrote, "nucleotides tend to fill functions collectively, rather than individually." Sequences that are nonfunctional at the level of individual nucleotides may function at higher levels involving physical interactions.[10]

Because the distance between enhancers and promoters is a factor in gene regulation, Zuckerkandl wrote in 2002, "genomic distance per se—and, therefore, the mass of intervening nucleotides—can have functional effects." He concluded: "Given the scale dependence of nucleotide function, large amounts of 'junk DNA,' contrary to common belief, must be assumed to contribute to the complexity of gene interaction systems and of organisms."[11] In 2007, Zuckerkandl (with Giacomo Cavalli) wrote that "SINEs and LINEs, which have been considered 'junk DNA,' are among the repeat sequences that would appear liable to have teleregulatory effects on the function of a nearby promoter, through changes in their numbers and distribution."[12]

Since enhancers can be tens of thousands of nucleotides away from the genes they regulate, bringing together enhancers and promoters that are on the same chromosome requires chromosome "looping."[13-17] The farther away an enhancer is from its promoter, the larger the loop must be, and the size of a loop depends on the length of the DNA. For physical reasons, a loop consisting only of DNA must be at least 500 nucleotides long, while a loop consisting of chromatin (because of its greater stiffness) must be at least 10,000 nucleotides long.[18] In such cases it is the sheer length of the DNA that matters, not whether it encodes RNAs.

In 2010, an international team of scientists reported that a long non-protein-coding RNA called HOTAIR[19] provides a "scaffold" for two molecular complexes involved in embryo development. HOTAIR consists of 2,146 RNA subunits; 300 at one end bind to the first complex, and 646 at the other end bind to the second. The intervening non-protein-coding subunits (all encoded by DNA) function by tethering the two complexes together at the proper distance from each other.[20]

The Second Level: Chromatin Organization

BECAUSE DNA is packaged into chromatin, and because RNA polymerase must have access to the DNA to transcribe it, the structure of chromatin is all-important in gene regulation. In many cases, various proteins and RNAs mediate the attachment of RNA polymerase to the DNA by interacting with specific sequences of nucleotides, but in some cases a mere change in the conformation (i.e., the three-dimensional shape) of chromatin can activate transcription by exposing the DNA to RNA polymerase.[21]

In 2007, scientists in Massachusetts produced a genome-scale, high-resolution three-dimensional map of DNA and found similar conformations that were independent of the underlying nucleotide sequences. They concluded that "considerably different DNA sequences can share a common structure," and they proposed that some transcription factors may be "conformation-specific… rather than DNA sequence-specific."[22]

Two years later, scientists reported that functional non-protein-coding regions of the human genome are correlated with "local DNA topography" that can be independent of the underlying sequence. "Although similar sequences often adopt similar structures," they wrote, "divergent nucleotide sequences can have similar local structures," suggesting that "they may perform similar biological functions." The authors of the report concluded that "some of the functional information in the noncoding portion of the genome is conferred by DNA structure as well as by the nucleotide sequence."[23]

Non-protein-coding RNAs contribute to chromatin structure. In many cases they do this by interacting with the DNA in a sequence-specific manner, but some RNAs may serve a mechanical role. In 2007, Spanish molecular biologists reported a "general structural role for RNA in eukaryotic chromatin." They found that RNA constitutes 2%-5% of purified chromatin and "contributes to its structural organization."[24]

The clearest example of a chromatin-level function that is independent of the exact DNA sequence is the "centromere," a special region on a eukaryotic chromosome that serves as the chromosome's point of attachment to other structures in the cell.

Centromeres

BEFORE A eukaryotic cell divides it makes a duplicate of each chromosome, and the duplicate copies of each chromosome are joined together at their centromeres. On the outward-facing surface of each centromere is a "kinetochore," which provides the point of attachment for microtubules that pull the duplicate chromosomes apart when the cell divides. (**Figure 7.2**)

The kinetochore is not simply a point of attachment. It is a complex structure composed of scores of different molecules, and it actively participates in moving chromosomes apart during cell division.[25-30] Yet it can form only on the foundation provided by the centromere.

Centromeres, in turn, can form only on the foundation provided by the chromosome. Yet centromeres are built upon long stretches of repetitive DNA that some biologists have regarded as junk.[31] Although much of the DNA that underlies centromeres is now known to be transcribed into RNAs that perform a variety of functions,[32-51] it turns out that centromere formation is to a great extent independent of the exact DNA sequence.

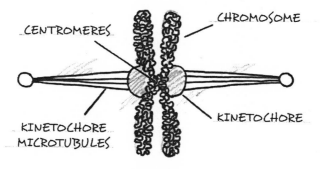

CENTROMERES

CHROMOSOME

KINETOCHORE

KINETOCHORE
MICROTUBULES

Figure 7.2 *Centromeres and kinetochores.* During cell division, duplicated chromosomes are joined by their centromeres. The gray bulge on the outward-facing surface of each centromere is a kinetochore, the attachment site for microtubules extending between the chromosome and a pole of the cell division apparatus. As the cell divides the duplicate chromosomes separate and are pulled to opposite poles by their kinetochore microtubules.

The DNA sequences of centromere regions vary significantly from species to species, though all centromeres function similarly.[52] If the chromosome region containing a centromere is artificially deleted and replaced by synthetic repetitive DNA, a functional centromere can form again at the same site.[53] Extra centromeres (called "neocentromeres") can also form abnormally elsewhere on a chromosome that already has one, or on a chromosome fragment that has separated from the part bearing a centromere.[54–55] It seems that centromeres (and their accompanying kinetochores) can form at many different places on a chromosome, regardless of the underlying DNA sequence. Yet the underlying chromatin must have certain characteristics that make centromere formation possible.

In the 1980s, biologists identified several proteins associated with centromeres and called them CENPs (for **CEN**tromere Proteins).[56–57] Subsequent research revealed that one of these, CENP-A, takes the place of some of the histones in chromatin.[58–60] The incorporation of CENP-A makes chromatin stiffer and provides a foundation for assembling the other components of centromeres and kinetochores.[61–62] In fact, centromeres in all organisms are associated with CENP-A, which must be present for a centromere and kinetochore to form, though CENP-A by itself is not sufficient.[63-64]

The modification of chromatin by CENP-A and other centromere-specific proteins can be passed down from generation to generation. Indeed, the location of a centromere on a particular chromosome can persist for thousands of generations. This sort of inheritance is called "epigenetic," meaning "on top of the genes," because it does not involve changes in the DNA sequence itself. From the perspective of the Central Dogma that DNA sequences determine the essential features of organisms by encoding proteins, centromeres are an enigma because they show that a cell can impose an essential but heritable structure on its DNA that is independent of the nucleotide sequence.

Although centromeric DNA sequences can vary significantly from species to species, there is evidence that some aspects of the DNA sequence are conserved.[65-66] In humans and other primates, centromere activity is normally associated with repeated blocks of 171 nucleotide subunits termed alpha-satellite DNA. (As we saw in Chapter 3, researchers in the 1960s discovered that a fraction of DNA consisting of millions of short, repeated nucleotide sequences produced "satellite" bands when DNA was centrifuged to separate it into fractions with different densities.) Every normal human centromere is located on alpha-satellite DNA.[67-70]

In 2002 and 2003, American biologists used alpha-satellite DNA from three different sources to make human artificial chromosomes and found that the results varied. They concluded "that centromere specification is at least partly dependent on DNA sequence."[71-72] Centromeres in the plant *Arabidopsis* (Chapter 5) are based on blocks of 178 nucleotide subunits with sequences that are completely different from alpha-satellite DNA, yet they are organized in the same way.[73-75]

But human neocentromeres form on parts of a chromosome that do *not* consist of alpha-satellite DNA, though the neocentromere DNA still has special characteristics—most notably, an unusually high proportion of LINEs.[76] These retrotransposons apparently play a role in localizing the CENP-A that is required for the formation of the centromere and kinetochore.[77-78] So centromere DNA must have certain characteristics, but it does not need to have a specific nucleotide sequence.

The Third Level: Chromosome Arrangement in the Nucleus

BETWEEN CELL divisions, chromosomes are not randomly distributed in the nucleus. Instead, they occupy distinct domains[79-82] that affect gene regulation—in part, by bringing together specific regions of the chromosomes and facilitating interactions among them.[83-88] Different cell and tissue types in the same animal can have different three-dimensional patterns of chromosomes in their nuclei, which account for at least some differences in gene expression.[89-90]

One notable feature of nuclear domains is their radial arrangement.[91] In 1998, biologists in New York reported that chromatin localized to the periphery of the nucleus in yeast cells tends to be "transcriptionally silent."[92] In 2001, British biologists wrote that "most gene-rich chromosomes concentrate at the centre of the nucleus, whereas the more gene-poor chromosomes are located towards the nuclear periphery."[93] In 2008, Dutch biologists reported that human chromosome domains associated with the periphery of the nucleus "represent a repressive chromatin environment."[94] The same year, several teams of researchers reported independently that they could suppress the expression of specific genes by relocating them to the nuclear periphery.[95-97]

These data are consistent with the observation that in most nuclei the gene-rich euchromatin is concentrated near the center while the gene-poor heterochromatin is situated more peripherally. Many factors might be involved in producing this radial arrangement, though biophysicists have proposed that one factor may be a tendency to establish a minimum-energy conformation that is independent of the exact sequence of nucleotides.[98-99]

Until recently, the only known exceptions to this radial arrangement occurred in some single-celled organisms,[100] but another newly discovered exception points to an important function of non-protein-coding

DNA that operates at the level of nuclear organization but is unrelated to the precise DNA sequence.

Non-Protein-Coding DNA Can Function as a Lens

THE RETINA of the vertebrate eye contains several different kinds of light-sensing cells. Cone cells detect colors and function best in bright light; rod cells are more numerous and more sensitive to low light. Nocturnal animals such as mice need to see under conditions of almost no light, so they need exceptionally sensitive rod cells. In 1979, medical researchers examined mouse retinas with an electron microscope and found that the heterochromatin in cone cells was located near the periphery of the nucleus (like most other eukaryotic cells), but in rod cells the heterochromatin was concentrated in "one large, central clump."[101] (**Figure 7.3**)

Figure 7.3 *Chromatin arrangement in the nucleus.* (Left) A simplified view of the arrangement of chromatin in most eukaryotic nuclei. Gene-poor heterochromatin (black) is on the periphery, and the gene content of the chromatin increases toward the center, which consists of gene-rich euchromatin (white). (Right) A simplified view of the inverted chromatin arrangement found in the nuclei of rod cells in the retinas of nocturnal mammals. Gene-rich euchromatin is on the periphery, while gene-poor heterochromatin is in the center. The centrally located heterochromatin acts as a liquid-crystal lens that focuses the few photons available at night onto the light-sensitive outer segments of the rod cells.

Another team of medical researchers used mice to study the genetic mutation responsible for an inherited human disease that causes nerve degeneration.[102] The team found that the mutation causes blindness in mice by altering the arrangement of the chromatin in rod cells. Instead of containing "a single, large clump of heterochromatin surrounded by a spare rim of euchromatin," the rods cell in mutant mice "showed a dramatic chromatin decondensation" and "resembled cone nuclei."[103]

Clearly, the unique localization of heterochromatin in the center of rod cell nuclei in the mouse retina is essential for normal vision in these animals. In 2009, European scientists called the unusual pattern of centrally located heterochromatin "inverted," and they reported finding an inverted pattern in the rod cell nuclei of various other animals that are primarily nocturnal (including cats, rats, foxes, opossums, rabbits and several species of bats) but not of animals that are primarily active in daylight (such as cows, pigs, donkeys, horses, squirrels and chipmunks). These scientists observed that the centrally located heterochromatin had a high refractive index—a characteristic of optical lenses—and by using a two-dimensional computer simulation they showed that a main consequence of the inverted pattern was to focus light on the light-sensitive segments of rod cells.[104–105]

In 2010, molecular biologists in France reported that the organization of the central heterochromatin in the rod nuclei of nocturnal mammals is consistent with a "liquid crystal model,"[106] and British biophysicists improved upon the 2009 study by using a new computer simulation to show that "the focusing of light by inverted nuclei" in three dimensions is "at least three times as strong" as it is in two.[107]

So at all three levels of the genomic hierarchy, there is evidence for functions that are independent of the exact DNA (or RNA) sequence. Like the evidence for sequence-dependent functions, the evidence for sequence-independent functions is almost certain to grow as scientists continue to expand their research horizon beyond the limits of the Central Dogma. There is a lot more to the genome (not to mention the living cell) than the protein-coding sequences in DNA.

Unfortunately, as we shall see in the next chapter, this fact has not prevented some recent apologists for Darwinism from trying to breathe new life into the myth of junk DNA.

8.

Some Recent Defenders
of Junk DNA

Dawkins, Miller, Futuyma, Collins, Coyne, and Avise are not the only biologists who still defend the notion of junk DNA. Since 2006 a number of other biologists have risen to its defense.

As we saw in Chapter 5, a team of Japanese and American biologists reported in 2003 that RNAs transcribed from a pseudogene increased the expression of the corresponding gene by serving as decoys for molecules that would otherwise degrade messenger RNAs transcribed from the gene itself.[1] In 2006, American biologists Todd Gray, Alison Wilson, Patrick Fortin and Robert Nicholls published a study that they claimed invalidated the 2003 report. If they had stopped with that claim, their article would simply have been a normal part of the scientific enterprise, in which all conclusions are subject to testing and, potentially, invalidation. But Gray and his colleagues went much further. After pointing out that ID advocates had written in a lay periodical that the 2003 report attested to "a purpose for junk DNA" and "even intelligent design,"[2] Gray and his colleagues wrote: "Each of these unlikely scenarios is now shown by our work to be incorrect." They concluded: "Our work reestablishes the evolutionary paradigm supported by overwhelming evidence that mammalian pseudogenes are indeed inactive gene relics."[3]

Yet even if the results published by Gray and his colleagues were valid, their conclusion would not logically follow. Invalidating a report of one function in one pseudogene cannot exclude other possible functions in that pseudogene, much less possible functions in other pseudogenes. Indeed, as we saw in Chapter 5, widespread transcription and sequence conservation imply that many pseudogenes are functional, and there is good evidence for specific functions in several cases—including two cases that are very similar to the one reported in 2003.[4-5] The sweeping pro-Darwin, anti-ID conclusion by Gray and his colleagues was obviously motivated by something other than evidence or logic—that is, by something other than science.

Genomic Dark Matter

IN 2009, University of Toronto biologist Timothy Hughes and his postdoctoral researcher Harm van Bakel published a scientific article challenging the notion that much of our DNA is transcribed into functional RNAs. Others had already used the term "dark matter" (borrowed from physics) to refer to non-protein-coding DNA[6-7] and the RNAs transcribed from it.[8] Hughes and van Bakel suggested that "the total volume of 'dark matter' transcription compared to the total transcriptional output of the genome may be smaller than initially estimated," and that "the functional role of most 'dark matter' non-coding RNAs remains unclear."[9]

In 2010, Hughes and van Bakel joined with two other University of Toronto researchers to publish an article concluding that "most 'dark matter' transcripts are associated with known genes" and that "the genome is not as pervasively transcribed as previously reported."[10] Hughes and his colleagues thereby directly contradicted a 2007 report that the ENCODE Project had found "convincing evidence that the genome is pervasively transcribed, such that the majority of its bases can be found in primary transcripts, including non-protein-coding transcripts."[11]

In a commentary based on the 2010 article by Hughes and his colleagues, science writer Richard Robinson concluded that their work "shows that most dark matter transcripts are likely to be by-products of transcription of known genes and that many of the rest of them are likely not messages of great import, but simple background noise."[12] And science writer Carl Zimmer reported on a blog affiliated with *Discover Magazine* that Hughes and his colleagues "used new methods to survey the RNA produced by the genome and compared their results to the ones from older methods. They found that most of their RNA came from regions of the genome that are already known to be protein-coding genes. Very little RNA came from elsewhere in the genome. They argue that the older methods were crude, so studies based on them were loaded with false positives."[13]

But Robinson and Zimmer should have checked the "Materials and Methods" section of the article. Hughes and his colleagues considered only "singleton" RNAs that "could be unequivocally mapped to unique positions in the genome," and they used a software program called "RepeatMasker" to discard the rest. They thereby biased their sample against most transcripts from repetitive DNA. Yet as we saw in Chapter 6, about half of our genome consists of repetitive DNA. Indeed, the official description of RepeatMasker states: "On average, almost 50% of a human genomic DNA sequence currently will be masked by the program."[14]

In the fraction they did analyze, Hughes and his colleagues based their results "primarily on analysis of PolyA$^+$ enriched RNA"—sequences that have a long tail consisting of many repeats of the DNA subunit containing adenine (A). Yet molecular biologists reported in 2005 that transcripts lacking the long tail (called PolyA$^-$ sequences) are twice as abundant in humans as PolyA$^+$ transcripts.[15] So Hughes and his colleagues not only excluded half of the human genome with RepeatMasker, but they also ignored two thirds of the RNA in the remaining half. It is no wonder that they found far fewer transcripts than have been found by the hundreds of other scientists who have been studying the human genome.

Ignoring their obvious methodological bias, Darwinian biologist (and outspoken atheist) P. Z. Myers praised the work of Hughes and his colleagues. According to Myers, "creationists" liked earlier reports of widespread functions in non-protein-coding DNA because "they detest the idea of junk DNA—that the gods would scatter wasteful garbage throughout our precious genome by intent was unthinkable, so any hint that it might actually do something useful is enthusiastically seized upon as evidence of purposeful design." Confessing that he himself falls "into the 'it's all junk' end of the spectrum," Myers welcomed the Toronto researchers' conclusions: "Well, score one for the more cautious scientists, and give the creationists another big fat zero... A new paper has come out that analyzes transcripts from the human genome using a new technique, and, uh-oh, it looks like most of the early reports of ubiquitous transcription were wrong." The bottom line, Myers concluded, is that "the genome is mostly dead, transcriptionally. The junk is still junk."[16]

So Myers, like Robinson and Zimmer, did not bother to look at the methodology used by Hughes and his colleagues—a methodology guaranteeing that their results would appear to support the myth of junk DNA.

Following publication of the 2010 article by Hughes and his colleagues, an international team of scientists reaffirmed earlier reports that RNAs "whose function and/or structure we do not understand (the so called 'dark matter' RNAs)" can constitute the majority of nuclear DNA-encoded, non-ribosomal RNA in a cell, and "a significant fraction arises from numerous very long, intergenic transcribed regions." The team sharply criticized Hughes and his colleagues for focusing "only on PolyA-selected RNA, a method that enriches for protein coding RNAs and at the same time discards the vast majority of RNA prior to analysis"—a method that is "certain to leave gaping holes in [our] understanding of the transcriptome."[17]

Despite this rebuttal of Hughes and his colleagues, Scottish evolutionary biologist Mark Blaxter perpetuated the myth of junk DNA in a December 2010 commentary in Science. Blaxter wrote: "Only 1% of the human genome is transcribed into protein-coding messenger RNA (mRNA) and non–protein-coding RNA (ncRNA), and DNA elements that control the expression of genes occupy another ~0.5%, suggesting that the remaining 'dark genome' is nonfunctional padding."[18] Blaxter did not cite any evidence for his 1% claim, which clearly contradicts the findings of many genome researchers.[19-20] Blaxter also contradicted an essay published in Science the week before, which surveyed the work of some of those genome researchers and reported that "about 80% of the cell's DNA showed signs of being transcribed into RNA."[21]

The Onion Test

IN 2007, Canadian biologist T. Ryan Gregory wrote: "Some non-coding DNA is proving to be functional, but this is still a minority of the non-coding DNA, and there is always the issue of the onion test when considering all non-coding DNA to be functional."[22] The "onion test," according to Gregory, "is a simple reality check for anyone who thinks they have come up with a universal function for non-coding DNA. Whatever your proposed function, ask yourself this question: Can I explain why an onion needs about five times more non-coding DNA for this function than a human?"[23]

The difference between the DNA content of an onion cell and that of a human cell is one piece of a larger puzzle called the "C-value paradox" or "C-value enigma."[24-30] Biologists have long known that the DNA content (the "C-value") of eukaryotic cells varies by a factor of several thousand, with no apparent correlation to organismal complexity or to the number of protein-coding genes. There is a strong positive correlation, however, between the amount of DNA and the volume of a cell and its nucleus—which affects the rate of cell growth and division.[31-32] Furthermore, in mammals there is a negative correlation between genome size and the rate of metabolism.[33] Bats have very high metabolic rates and relatively small genomes.[34-35] In birds, there is a negative correlation between C-value and resting metabolic rate.[36-37] In salamanders, there is also a negative correlation between genome size and the rate of limb regeneration.[38]

Gregory has written extensively on the C-value enigma,[39-42] and various hypotheses have been proposed to explain it.[43-48] One of those hypotheses attempts to explain the enigma by the accumulation of "junk DNA" or "selfish DNA," but—as Gregory himself has pointed out—that explanation cannot make sense of the correlations noted above.[49] "Under the traditional junk DNA and selfish DNA theories,"

Gregory wrote in 2005, "the relationship between genome size and cell size is considered purely coincidental." Since this approach is incapable of explaining the correlation between C-value and cell size, "the strictly coincidental interpretation has been rejected."[50]

But if Gregory rejects the accumulation of "junk DNA" as an explanation for the C-value enigma, why does he use the "onion test" to defend the notion that most non-protein-coding DNA is nonfunctional? Something peculiar is going on here. Let's take a closer look at his reasoning.

First, Gregory directs his challenge to "anyone who thinks they have come up with a universal function for non-coding DNA." Yet there probably is no such person. As we have seen, scientists know of many functions for non-protein-coding DNA. Nobody claims that there is "a universal function" that applies both to mammals and to onions. Based on the evidence, scientists have proposed that non-protein-coding intronic DNA helps to regulate alternative splicing in brain cells, and that non-protein-coding repetitive DNA plays a role in placental development. Why should those scientists justify their proposals by referring to onions, which have neither brains nor placentas?

Second, Gregory makes it clear that his true goal is to defend Darwinian evolution and attack intelligent design. One way he does this is by misrepresenting the latter. The same year he proposed the onion test he wrote that in order for ID to be considered scientific its proponents must "specify the basis for assuming that all non-coding DNA must be functional."[51] But ID proponents do *not* assume that all non-coding DNA must be functional. They infer that it is unlikely that most of our DNA would be nonfunctional; therefore, scientists should continue looking for functions.[52-53]

Gregory misrepresents not only ID but also the logic of the argument. In 2007 he wrote: "It is commonly suggested by anti-evolutionists that recent discoveries of function in non-coding DNA support intelligent design and refute 'Darwinism.'"[54] But Dawkins, Futuyma, Shermer, Collins, Kitcher, Miller, Coyne, and Avise argue exactly the opposite: They all claim that non-protein-coding DNA supports Darwinism and refutes intelligent design. It is *their* claim that is the issue here—and "recent discoveries of function in non-coding DNA" refute it. Gregory stands the debate on its head.

So the onion test is a red herring. Why onion cells have five times as much DNA as human cells is an interesting question, but it poses no challenge to the growing evidence against the myth of junk DNA.

9.

SUMMARY OF THE CASE FOR

FUNCTIONALITY IN JUNK DNA

OST OF OUR DNA DOES NOT CODE FOR PROTEINS; ON THAT, EVeryone agrees. The question here is whether non-protein-coding DNA is nonfunctional "junk" that provides evidence for Darwinian evolution and against intelligent design.

Evidence for the functionality of non-protein-coding DNA falls into two broad categories: The first consists of evidence suggesting that such DNA is *probably* functional. This evidence comes from two sources; the first source is the transcription of most non-protein-coding DNA into various RNAs. If only protein-coding regions of DNA were functional, then organisms that are struggling to survive probably wouldn't waste precious energy transcribing non-protein-coding regions into useless RNAs. Yet as we saw in Chapter 3, organisms transcribe most of their DNA into RNA, suggesting that non-protein-coding DNA is probably functional.

A second source of evidence in the first category comes from comparisons of DNA sequences in different organisms. According to evolutionary theory, different lineages inherit their DNA from a common ancestor. If two lineages inherit non-protein-coding DNA that is nonfunctional, it will be unaffected by natural selection and tend to accumulate mutations in a random manner. Many generations later, the non-protein-coding DNA in the two descendant lineages will be very different. On the other hand, if the non-protein-coding DNA is functional, natural selection will tend to weed out mutations. In evolutionary terminology, the descendants' non-protein-coding sequences will be "conserved."

Turning the logic around, evolutionary theory implies that if evolutionarily divergent organisms share similar non-protein-coding DNA sequences, those sequences are probably functional. As we have seen, many non-protein-coding DNA sequences are conserved, suggesting that they serve biological functions.

So in the first category, widespread transcription and sequence conservation suggest that much "junk" DNA is probably functional, though they do not tell us what the precise functions are. The second broad category consists of evidence for specific biological functions of non-protein-coding DNA. The first category was discussed in Chapter 3, and the second category was discussed in Chapters 4–7.

Chapter 3

RNAs TRANSCRIBED from non-protein-coding DNA play significant roles in controlling whether, where, and to what extent the protein-coding regions are transcribed. Non-protein-coding RNAs are also involved in regulating the translation of RNAs into proteins. The process by which a DNA sequence yields a functional product (such as a protein) is called "gene expression."

In 2006, Spanish scientists reported that non-protein-coding RNAs "regulate virtually all aspects of the gene expression pathway, with profound biological consequences."[1] In 2009, biologists in Japan noted that since "research in the recent few years has identified an unexpectedly rich variety of mechanisms by which non-coding RNAs act," it is likely "that we have identified probably only a few of the many potential functional mechanisms" of non-protein-coding RNAs.[2]

Recent discoveries show that non-protein-coding RNAs are essential constituents of "paraspeckles," domains within the nucleus that play a role in gene expression. By binding to certain proteins, the non-protein-coding RNAs help to stabilize the structure of paraspeckles so they can persist through cell divisions even though they are not bounded by membranes.

Chapter 4

GENES IN eukaryotes (cells with nuclei) are divided into protein-coding "exons" and non-protein-coding "introns." Exons and introns are both transcribed, but the latter are then cut out and the former are spliced together in alternative ways. As a result, a single protein-coding region of DNA can give rise to hundreds or thousands of different proteins.

Yet introns are not just passive spacers: A team of Canadian and British scientists studying splicing codes in mouse tissues reported in 2010 that introns are rich in splicing-factor recognition sites. It had previously been assumed that such sites tend to be close to the affected exons, but the team concluded that their results suggested "regulatory elements that are deeper into introns than previously appreciated."[3]

In humans, introns also encode a majority of the small RNAs involved in the molecular machinery that translates messenger RNAs into proteins. In addition, non-protein-coding RNAs from introns influence gene expression by modifying chromatin—the complex combination of DNA, RNAs and proteins that makes up chromosomes.

Chapter 5

A PSEUDOGENE is a DNA sequence that appears to be an inactive copy of a sequence that elsewhere (or in another organism) codes for protein. But some presumed pseudogenes have turned out to produce functional proteins, and thus are not pseudogenes at all.

Some other pseudogenes produce RNAs that suppress the expression of their corresponding functional genes. DNA consists of two complementary strands; biologists use to think that only one (the "sense strand") is transcribed into RNA, while the second ("the antisense strand") functions only as a copying template during DNA replication. It is now known, however, that RNAs are produced from both strands. Thus pseudogene DNA can be transcribed into a non-protein-coding RNA that is complementary to the protein-coding RNA from the functional gene. The former can bind to the latter and thereby inactivate it—a process known as "RNA interference."

Still other pseudogenes produce RNAs that *increase* the expression of their corresponding functional genes. The cell contains molecules that control the level of gene expression by degrading protein-coding RNAs after they have been translated into protein. Although the RNA transcribed from the pseudogene does not code for protein, it is otherwise very similar to the RNA transcribed from the protein-coding gene. Thus the former can take the place of the latter in the presence of RNA-degrading molecules, leaving the protein-coding RNA free to continue making protein. In the words of some American biologists who studied this phenomenon, pseudogene RNAs serve as "perfect decoys."[4]

Chapter 6

ABOUT HALF of the human genome consists of repetitive non-protein-coding DNA. Most of this repetitive DNA consists of Long Interspersed Nuclear Elements ("LINEs") and Short Interspersed Nuclear Elements ("SINEs"). Some other repetitive DNA elements look as though they were derived from RNA viruses and thus are called "endogenous retroviruses" ("ERVs"). There is growing evidence that these (and other) categories of repetitive non-protein-coding DNA perform various functions.

For example, female mammals have two X chromosomes, one of which must be inactivated for an embryo to develop normally. In 2010, biologists reported that LINEs function at two different levels to produce X chromosome inactivation: First, LINE DNA produces a rearrangement in the chromatin that inactivates some genes; second, RNAs transcribed from LINEs coat and silence other portions of the X chromosome. LINEs also participate in repairing DNA breaks, mobilizing various other RNAs within the cell, and regulating genes that are expressed differently in fetuses and adults.

SINEs help to regulate the transcription of DNA into RNAs, the alternative splicing of RNAs, and the translation of RNAs into proteins. In 2009, Colorado scientists reported evidence that "SINE-encoded RNAs indeed have biological functions," and they concluded that the evidence "has refuted the historical notion that SINEs are merely 'junk

DNA."[5] SINEs also influence transcription by affecting chromatin. In 2007, biologists provided evidence from mouse embryos suggesting that tissue-specific transcription of SINEs "may represent a developmental strategy for the establishment of functionally distinct domains within the mammalian genome to control gene activation."[6]

ERVs help to regulate human genes involved in producing blood cells, transporting bicarbonate, and metabolizing fat. ERVs also regulate gene expression in the gastrointestinal tract, mammary glands, and testes. Probably the best-studied function of ERVs, however, is in the placenta. When an early mammalian embryo implants itself in the wall of the uterus, cells from the embryo migrate into the uterine wall and then fuse into a single multinucleated cell to facilitate rapid transfer of nutrients from the mother to the fetus. The all-important fusion of those cells requires an ERV-derived protein called "syncytin" (pronounced sin-SIGHT-in). In 2009, a British scientist wrote that it used to be "an open question" whether ERVs "simply represented junk or selfish DNA," but he called syncytin "compelling evidence" that at least some ERVs are making "a specific contribution to normal physiology.[7]

Chapter 7

CHAPTERS 3–6 describe functions of so-called "junk DNA" that depend on RNAs with sequences that regulate gene expression or perform other important roles in living cells. There remain vast stretches of DNA for which no sequence-dependent functions have been identified, but some of those vast stretches have other roles.

The genome is hierarchical, and it functions at three levels: the DNA molecule itself; the DNA-RNA-protein complex that makes up chromatin; and the three-dimensional arrangement of chromosomes in the nucleus. At all three of these levels, DNA can function in ways that are independent of its exact nucleotide sequence.

At the first level, some biologists have argued that DNA sequences can affect gene expression simply by their length. Molecular biologist Emile Zuckerkandl wrote in 2002 that "genomic distance per se—and, therefore, the mass of intervening nucleotides—can have functional effects." Thus "large amounts of 'junk DNA,' contrary to common belief, must be assumed to contribute to the complexity of gene interaction systems and of organisms."[8] For example, the sheer length of introns could affect the rate of transcription. Length could also affect the size of loops that enable distant parts of the DNA to interact, or the size of non-protein-coding RNAs that tether regulatory molecules at appropriate distances from each other.

At the second level, chromatin structure profoundly affects gene expression, but chromatin structure is in some places independent of the underlying DNA sequence. In 2009, scientists reported that "divergent nucleotide sequences can have similar local structures," suggesting that "they may perform similar biological functions." The scientists concluded that "some of the functional information in the non-coding portion of the genome is conferred by DNA structure as well as by the nucleotide sequence."[9]

The best-studied examples of sequence-independent chromatin function, however, are centromeres. A centromere is a special region on a eukaryotic chromosome that serves as the chromosome's point of attachment to other structures in the cell. The centromere also provides the foundation for the kinetochore, a complex molecular apparatus that moves chromosomes apart during cell division. Centromeres function similarly in all organisms, yet the DNA sequences underlying them differ significantly. What matters is not so much the nucleotide sequence as a set of centromere-specific molecules that the cell uses to package the chromatin in a particular way.

At the third level, the position of a chromosome inside the nucleus is important for gene regulation. In most cells, the gene-rich portions of chromosomes tend to be concentrated near the center of the nucleus, and a gene can be inactivated by artificially moving it to the periphery. In some cases, however, the pattern is inverted: Rod cells in the retinas of nocturnal mammals contain nuclei in which the non-protein-coding parts of chromosomes are concentrated near the center of the nucleus, where they form a liquid crystal that serves to focus dim rays of light.

Chapter 8

ALTHOUGH SCIENTISTS have discovered many functions for so-called "junk DNA," a few biologists (in addition to those cited in Chapter 2) have recently come to the defense of the notion. In 2007, Canadian biologist T. Ryan Gregory wrote: "Some non-coding DNA is proving to be functional, but this is still a minority of the non-coding DNA, and there is always the issue of the onion test when considering all non-coding DNA to be functional."[10] The onion test, according to Gregory, "is a simple reality check for anyone who thinks they have come up with a universal function for non-coding DNA. Whatever your proposed function, ask yourself this question: Can I explain why an onion needs about five times more non-coding DNA for this function than a human?"[11]

Yet no one claims to have come up with "a universal function for non-coding DNA." Instead, scientists have discovered many different functions for non-protein-coding DNA. Those functions include regulating alternative splicing in brain cells and playing an essential role in placental development. Why should the scientists who discovered those functions have to justify their work by referring to onions, which have neither brains nor placentas?

In 2010, some University of Toronto researchers reported that "the genome is not as pervasively transcribed as previously reported."[12] According to Darwinist (and atheist) P. Z. Myers, "creationists" liked earlier reports of widespread functions in non-protein-coding DNA because "they detest the idea of junk DNA—that the gods would scatter wasteful garbage throughout our precious genome by intent was unthinkable, so any hint that it might actually do something useful is enthusiastically seized upon as evidence of purposeful design." Myers welcomed the Toronto researchers' conclusions: "Well, score one for the more cautious scientists, and give the creationists another big fat zero... A new paper has come out that analyzes transcripts from the human genome using a new technique, and, uh-oh, it looks like most of the early reports of ubiquitous transcription were wrong." The bottom line, Myers concluded, is that "the genome is mostly dead, transcriptionally. The junk is still junk."[13]

But the Toronto researchers used methods that virtually guaranteed their results. They began by using a software program that excludes most repetitive DNA (which makes up half of the human genome), then they threw out about two-thirds of the RNAs from the remaining half. A rebuttal subsequently published by genome biologists criticized the Toronto researchers for discarding "the vast majority of RNA prior to analysis"—a method that is "certain to leave gaping holes in [our] understanding of the transcriptome."[14]

Given the abundant and growing evidence for functionality in non-protein-coding DNA, it seems that recent defenders of the myth of junk DNA—like the authors cited in Chapter 2—are motivated by something other than the scientific evidence.

10.

FROM JUNK DNA TO A NEW
UNDERSTANDING OF THE GENOME

IN CHAPTER 9 WE REVIEWED SOME OF THE EVIDENCE AGAINST THE myth of junk DNA presented in Chapters 3 through 8. In this chapter, we return to the arguments based on junk DNA that we encountered in Chapter 2. Richard Dawkins, Kenneth Miller, Michael Shermer, Francis Collins, Philip Kitcher, Jerry Coyne and John Avise all claimed that most of our DNA is nonfunctional junk, and that this provides evidence *for* Darwinian evolution and *against* intelligent design (ID).

To be fair, we should note that Collins acknowledged in 2006 that some repetitive DNA elements "may play important regulatory roles," but he dismissed this as a "small fraction" of the total.[1] And Avise wrote in 2010 that "several instances are known or suspected in which a pseudogene formerly assumed to be genomic 'junk' was later deemed to have a functional role in cells. But such cases are almost certainly exceptions rather than the rule."[2]

Futuyma acknowledged even more in 2005, when he wrote: "More than 10 percent of noncoding DNA is highly conserved…, suggesting a function," and "many noncoding regions, including introns, are transcribed into RNA sequences" that "perform important functions in gene regulation."[3] Nevertheless, Futuyma argued (like the others) that pseudogenes are nonfunctional junk, providing evidence for Darwinism and against ID.

Speaking for Science?

ALL EIGHT of the authors cited here present themselves as spokesmen for science. Yet science depends on evidence, and the tide of the evidence is clearly running against them. The previous chapters cite hundreds of published articles by over 1,000 scientists on 5 continents, but they are just a small sample. Anyone with a computer and an Internet connection can go to PubMed[4]—a freely accessible database of scientific articles maintained by the U. S. National Institutes of Health—and find hundreds of additional articles about the functions of non-protein-coding DNA. More are coming out every week.

Shermer and Kitcher are not scientists; perhaps they were just parroting what they heard from their scientific colleagues. But Shermer and Kitcher are scholars who presumably have computers and access to the Internet, so one might wonder why they didn't check the facts for themselves before buying into the myth of junk DNA.

Dawkins studied bird behavior in the 1960s, but since then he has spent his career writing popular books and articles defending Darwinism and preaching atheism. Obviously, he is out of touch with recent genomics research. Yet from 1995 to 2008 he was Professor for the Public Understanding of Science at Oxford. As such, he should have made at least some effort to familiarize himself with the evidence. Yet even now, he continues to defend the myth.

Coyne and Avise are professors of genetics at major universities, so they cannot claim ignorance of the genomic evidence without thereby admitting negligence or incompetence. In fact, one of Coyne's colleagues at the University of Chicago is James Shapiro, co-author of the 2005 article cited in Chapter 6 that listed over 80 known functions for non-protein-coding repetitive DNA.[5] But if Coyne and Avise were not ignorant of the evidence, then they misrepresented it—and they continue to do so. Like Dawkins, Shermer and Kitcher, they have forfeited any claim they might have had to be speaking for science.

Collins was head of the Human Genome Project from 1993 to 2007, so even before he published his *Language of God* in 2006 he should have been aware of the enormous amount of evidence being published on the functions of non-protein-coding DNA. In Collins's defense, however, it should be noted that he (unlike the others) subsequently recanted his belief in the myth of junk DNA. In 2007, he was a co-author of the EN-

CODE Project's landmark announcement that "the genome is pervasively transcribed."[6] He was also director of the National Human Genome Research Institute, which issued a press release at the time stating that the ENCODE Project's announcement "challenges the long-standing view that the human genome consists of a relatively small set of discrete genes, along with a vast amount of so-called junk DNA that is not biologically active."[7] Collins then declared in an interview for *Wired* magazine's blog that "I've stopped using the term" junk DNA.[8]

In 2010 Collins published another book, *The Language of Life*, in which he wrote that the "discoveries of the past decade, little known to most of the public, have completely overturned much of what used to be taught in high school biology. If you thought the DNA molecule comprised thousands of genes but far more 'junk DNA,' think again." Although he continued to maintain that our genome is "littered with repetitive sequences," of which only "a small fraction" are known to be useful, Collins acknowledged that "some DNA we used to call 'junk' is useful."[9]

Indeed, he concluded, "only about 1.5 percent of the human genome is involved in coding for protein," but "that doesn't mean the rest is 'junk DNA.' A number of exciting new discoveries about the human genome should remind us not to become complacent in our understanding of this marvelous instruction book. For instance, it has recently become clear that there is a whole family of RNA molecules that do not code for protein. These so-called non-coding RNAs are capable of carrying out a host of important functions, including modifying the efficiency by which other RNAs are translated. In addition, our understanding of how genes are regulated is undergoing dramatic revision, as the signals embedded in the DNA molecule and the proteins that bind to them are rapidly being elucidated. The complexity of this network of regulatory information is truly mind-blowing."[10]

Apparently, however, Collins's followers have not gotten the memo. In 2007, Collins founded The BioLogos Foundation to promote his view that "once life arose, the process of evolution and natural selection permitted the development of biological diversity and complexity over very long periods of time. Once evolution got under way, no special supernatural intervention was required."[11] When he was appointed Director of the U. S. National Institutes of Health in 2009, Collins handed over the leadership of the foundation to biologist Darrel Falk and science and religion scholar Karl Giberson,[12] both of whom still rely on junk DNA to argue against intelligent design.

In March 2010, after claiming (falsely, as we saw in Chapter 8) that ID "predicts that the DNA in the human genome (and other organisms) is fully functional," Falk wrote that although "plenty of magnificent 'sense' is scattered throughout the genome, coding for absolutely marvelous things," yet "this still doesn't negate the fact that almost certainly much, if not most, of the DNA plays no role."[13] The same month, Giberson wrote, "If we say that an intelligent agent has produced certain strings of DNA," then "what about DNA strings that look like gibberish? Why did our intelligent agent produce an information-rich string and sandwich it between two pieces of nonsense?"[14] If Collins has repudiated the myth of junk DNA, why do his followers at The BioLogos Foundation continue to promote it?

How Darwinists Might Respond

ALTHOUGH THE tide of evidence is running against the myth of junk DNA, some biologists (as we saw in Chapter 8) have made scientific claims that seem at first glance to support it. Now, in response to this book, some Darwinists might fall back on a tactic they used a few years ago—one that is based on misrepresentation and intimidation.

The National Center for Science Education (NCSE) is a pro-Darwin lobby group that aggressively opposes creationism, intelligent design, and even scientific criticisms of Darwinism in biology classrooms. In 2002, the pro-ID Discovery Institute published summaries of 44 articles in scientific journals and books that "represent dissenting viewpoints that challenge one or another aspect of neo-Darwinism (the prevailing theory of evolution taught in biology textbooks), discuss problems that evolutionary theory faces, or suggest important new lines of evidence that biology must consider when explaining origins."[15] The NCSE then contacted the authors of the articles to ask whether they "considered their work to provide scientific evidence for intelligent design" or "considered their work to provide scientific evidence against evolution."[16-17]

Of course, the Discovery Institute never claimed that the 44 articles provided "scientific evidence for intelligent design" or "scientific evidence against evolution" (which, as we saw in Chapter 1, can mean many things). Nevertheless, the NCSE's misleading questionnaire evoked angry responses from some of the articles' authors, who were understandably indignant at the suggestion that they were pro-ID or anti-evolution.[18]

It's possible that the NCSE or others might resort to the same deceptive and intimidating tactic again in response to this book. So I want to make myself very clear: I am *not* claiming that the authors of articles I cite in this book on the functions of non-protein-coding DNA are pro-ID or anti-evolution. I argue only that their work provides evidence against the notion that most of our DNA is "junk."

Theology Masquerading as Science?

APART FROM the growing evidence for functions in non-protein-DNA, there is another problem with the arguments of Dawkins, Miller, Futuyma, Shermer, Collins, Kitcher, Coyne and Avise. In the books cited above, all eight of these authors rely on speculations about why a creator or designer would or would not have done certain things.

Dawkins and Collins (and Coyne, in his discussion of the vitamin C pseudogene; see the Appendix) explicitly mention a "Creator." Miller, Futuyma, Shermer and Coyne refer to a "designer" (whom Futuyma also calls "God"). Kitcher mentions an "Intelligence" whom ID commits to "a whimsical tolerance of bungled designs," and Avise refers to a "wise engineer" and a "caring cognitive agent." Regardless of the exact words they use, all eight authors speculate on the motives of this entity.

Intelligent design does not rely on such speculations. According to ID, it is possible to infer from evidence in nature that some features of the world, and of living things, are better explained by an intelligent cause than by unguided natural processes. If the evidence shows that a feature has characteristics (such as specified or irreducible complexity) that in our experience invariably originate in intelligence, then a design inference is warranted. Although design implies a designer—an intelligent agent—ID does not tell us whether the designer is "beneficent," "wise," "caring," or "whimsical"—much less a Creator (which classically means a supernatural being who creates from nothing).

Normally, science tests theories against evidence from nature. Why are these eight supposed spokesmen for science defending Darwinism with speculations about the motives of a designer? Actually, they are following in the footsteps of Charles Darwin himself. He called *The Origin of Species* "one long argument,"[19] and it took this general form: The facts of nature are "inexplicable on the theory of creation," but make sense on his theory of descent with modification.[20-21] Yet there is something odd about this manner of reasoning. Would a geologist argue for continental drift by asking, "Why, on the theory of creation, should the eastern contour of the Americas resemble the western contour of Europe and Africa?" Or would a physicist argue for a theory of gravity on the grounds that the fall of an apple is "inexplicable on the theory of creation?"

In 1979, Georgia State University historian Neal C. Gillespie noted that *The Origin of Species* was "significantly dependent on theology" for the force of its argument. "Darwin's theological defense of descent with modification" rested on his conception of the creator, and *The Origin of Species* "not only has numerous references to such a creator, but theological arguments based on such a conception had some importance in its overall logic."[22] According to biophysicist Cornelius G. Hunter, the essence of Darwin's "one long argument" was that "evolution is true because divine creation is false." Darwin started with an idea of "how God would go about creating the world" and found that it did not match the facts of nature, "but the mismatch depends every bit as much on the theology as on the science."[23]

Philosopher of biology Paul A. Nelson has observed that "the use by many biologists and philosophers of theological arguments for evolution" is a "remarkable but little studied aspect of current evolutionary theory."[24] Historian of science Gregory Radick summarizes the Darwinists' principal argument as follows: "No Designer worth His salt would have created" the features that we actually find in nature. "It would be hard to exaggerate the importance of this argument," Radick wrote, "from Darwin's day to our own, as a means of disqualifying the Designer explanation and making room for Darwinian descent with modification."[25-27]

Do arguments based on speculations about a creator or designer have a legitimate place in science? Not according to Canadian biologist Steven Scadding, who once wrote that although he accepted evolutionary theory, he objected to defending it on the grounds that a creator would or would not do certain things. "Whatever the validity of this theological claim," Scadding concluded, "it certainly cannot be defended as a scientific statement, and thus should be given no place in a scientific discussion of evolution."[28]

The Logic of the Argument

IF WE ignore their theological speculations, we can state the argument of our eight authors in the following simplified form:

- If most human DNA is junk, then Darwinism is true and ID is false;

- Most human DNA is junk;

- Therefore Darwinism is true and ID is false.

By the rules of classical logic, affirming the antecedent ("most human DNA is junk") establishes the truth of the consequent ("Darwinism is true and ID is false"). So if it were true that most human DNA is junk, this argument would logically establish the truth of Darwinism and the falsity of ID. It is not true, however, that most human DNA is junk. In light of the evidence, the argument of our eight authors logically tells us nothing about the truth or falsity of Darwinism or ID. All it tells us is that the writers have put their faith in a failed argument.

It would not help their argument to point out (correctly) that there is still much of our DNA for which no function is known, and that some of this might indeed turn out to be "junk." Saying that *some* of our DNA *might* be junk is very different from claiming that *most* of our DNA *is* junk—and that the latter provides evidence for Darwinism and against ID. Indeed, holding out for the nonfunctionality of large amounts of our DNA hardly seems like a promising strategy, given the rate at which new functions are being reported in the scientific literature. Junk DNA advocates have to retreat every time a new function is found. In effect, they are relying on an argument from ignorance—a sort of "Darwin of the Gaps"—that becomes less tenable with each new scientific discovery.[29]

Can the Genome Support a Design Inference?

THE MYTH of junk DNA is effectively dead. But most of the scientists whose work helped to bury it are not advocates of intelligent design, and refuting the myth of junk DNA is not the same as arguing that ID is true. So the question remains: Can recent genome evidence lead to an inference of design?

In 1994 Kenneth Miller wrote: "If the DNA of a human being or any other organism resembled a carefully constructed computer program, with neatly arranged and logically structured modules each written to fulfill a specific function, the evidence of intelligent design would be overwhelming."[30] Only a year later, computer programmer and Microsoft chairman Bill Gates wrote: "DNA is like a computer program but far, far more advanced than any software ever created."[31]

In 2004, ID theorist Stephen C. Meyer expanded upon Gates's statement. "Like meaningful sentences or lines of computer code," Meyer wrote, "genes and proteins are also *specified* with respect to function. Just as the meaning of a sentence depends upon the specific arrangement of the letters in a sentence, so too does the function of a gene sequence depend upon the specific arrangement of the nucleotide bases in a gene." DNA thereby "conveys information."[32] Meyer expanded this argument further in his 2009 book *Signature in the Cell.*[33]

As we have seen, however, there is growing evidence that protein-coding genes are not the only parts of DNA that function by virtue of specific nucleotide sequences. Much of what used to be considered junk also carries sequence-dependent biological information. As design theorist William A. Dembski wrote in 2004: "For years now evolutionary biologists have told us that the bulk of genomes is junk and that this is due to the sloppiness of the evolutionary process. That is now changing. For instance, researchers at the University of California at San Diego are finding that long stretches of seemingly barren DNA sequences may form a new class of noncoding RNA genes scattered, perhaps densely, throughout animal genomes. Design theorists should be at the forefront in unpacking the information contained within biological systems."[34]

Information theorists have written extensively about sequence-dependent information in linear DNA sequences.[35–37] Yet sequence-dependent biological information is not straightforwardly linear. Since a single protein-coding segment of DNA can be transcribed from multiple sites, and both the sense and antisense strands can be transcribed (**Figure 3.6**), some genes contain multiple codes. In 2007, an international team of genome researchers identified 40 human genes that probably have "overlapping coding regions," a feature that the researchers concluded "is nearly impossible by chance."[38] The same year, Israeli scientists noted that although many regulatory elements reside in non-protein-coding regions of the genome, genes also carry—in addition to the code for a protein—"parallel codes" that include "binding sequences for regulatory and structural proteins, signals for splicing, and RNA secondary structure." The Israeli scientists concluded that the specification of amino acids by three-nucleotide "codons" in DNA is "nearly optimal for allowing additional information within protein-coding sequences."[39]

Commenting on the Israelis' work, American scientists noted that embedding multiple codes in a single gene is like "sending secret messages that are 'camouflaged' in unsuspicious looking communications (steganography)"—a form of cryptography. The simultaneous communication of two written messages, one of which is embedded in the other, "is similar to that of providing a template for an amino acid sequence together with noncoding information in a nucleotide sequence."[40]

Three years earlier, Dembski had listed "biosteganography" as one possible source of evidence for intelligent design in biological systems. "If these systems are designed," he wrote in 2004, "we can expect the information to be densely packed and multilayered." Thus "dense, multilayered embedding of information is a prediction of intelligent design."[41] In 2010, biologists reported the embedding of complex information processing networks—a characteristic of very large scale integrated computer circuits—in the nervous systems of both humans and roundworms.[42]

Not all biological information is sequence-dependent. As we saw in Chapter 7, the genome functions at three levels (the DNA molecule, the organization of chromatin, and the position of chromosomes within the nucleus). At all three levels, there is evidence for functions that are independent of the specific nucleotide sequence. Do we need a broader concept of biological information to understand sequence-independent functions? And might those functions support a design inference?

Genome researcher Richard von Sternberg thinks so. He has analyzed the genome-as-computer metaphor in the light of recent evidence and concluded that we need a new model of the genome that goes far beyond the limitations of the Central Dogma and neo-Darwinian theory.[43-44] Sternberg gives several reasons for this. First, the information carried by nucleotide sequences—both protein-coding and non-protein-coding—is bidirectional, multilayered, and interleaved, rather than simply linear. Second, repetitive elements format and punctuate the genome at different scales, producing a multidimensional filing system.[45] Third, cells can write codes onto non-protein-coding DNA, as they do in the case of centromeres—so the phenotype is not reducible to the genotype.

Thus the Central Dogma ("DNA makes RNA makes protein makes us") is untenable. The genome is actually a multilevel computational device in which many of the operations occur as interactions among components—what Sternberg calls "metaprogramming." And contrary to neo-Darwinism, the DNA sequence is not simply a linear code that can be mutated indefinitely to generate new information. Instead, it is highly specified to function as one component of a multidimensional system.

Sternberg argues that intelligent design suggests the following hypothesis: The organization of DNA strings along the genome is optimized for the establishment of multidimensional codes at all scales, and each species has a unique and elaborately ordered arrangement of chromosome regions that maximizes the information its genome can carry. The hypothesis is scientific, because it entails two predictions that can be empirically falsified: The first is that the genome of one species cannot be transformed into the genome of another species by random re-arrangements, since this would compromise the formatting, indexing, and punctuation of DNA files. The second is that any observed chromosome changes that result in normal fitness will be those that maintain genomic optimization.

Where Do We Go From Here?

SCIENTISTS MAKE progress by testing hypotheses against the evidence. But when scientists ignore the evidence and cling to a hypothesis for philosophical or theological reasons, the hypothesis becomes a myth. Junk DNA is such a myth, and it's time to leave it behind—along with other discarded myths from the past.

As recent discoveries have demonstrated, we are just beginning to unravel the mysteries of the genome. Indeed, the same can be said of living organisms in general. But assuming that any feature of an organism has no function discourages further investigation. In this respect, the myth of junk DNA has been a science-stopper.

Not any more. For scientists willing to follow the evidence wherever it leads, these are exciting times.

APPENDIX:

THE VITAMIN C PSEUDOGENE

Vitamin C (ascorbic acid) is essential for many biochemical reactions in living cells. Yet we are unable to synthesize it in our bodies, so we need to supplement our diets with it. Guinea pigs, chimpanzees and several species of monkeys are also unable to synthesize their own vitamin C;[1-2] so are some (but not all) species of bats,[3-5] some (but not all) species of birds,[3,6-7] and some (but not all) species of fishes.[8-9]

Vitamin C synthesis requires four enzymes, of which we have three; our cells also contain a segment of DNA very similar to the gene for the fourth enzyme, L-gulonolactone γ-oxidase (abbreviated GULO or GLO), but this segment of DNA is not translated into protein.[10-11] In other words, the human genome includes a vitamin C pseudogene, *GLO*. (Gene names and abbreviations are customarily italicized, while the corresponding proteins are not.)

As we saw in Chapter 2, Brown University biologist Kenneth R. Miller and University of Chicago geneticist Jerry A. Coyne have argued that the *GLO* pseudogene provides evidence for Darwinian evolution—in particular, for the common ancestry of humans and other primates—and evidence against intelligent design or creation.

Kenneth Miller's Argument

"If the designer wanted us to be dependent on vitamin C," wrote Miller in 2008, "why didn't he just leave out the *GLO* gene from the plan for our genome? Why is its corpse still there?" Miller concedes that proponents of intelligent design could argue that the designer originally gave us a functional *GLO* gene, but it was later inactivated by mutations; the inactive pseudogene would then have been inherited by all living humans from their common ancestor.[12]

"But in that simple conclusion lies the undoing of any claim for our separate ancestry as a species," Miller continued, because humans are not the only species in which the GLO gene is broken. A vitamin C pseudogene is also found in "a certain group of primates, the very ones that happen to be our closest evolutionary relatives. Orangutans, gorillas, and chimps require vitamin C, as do some other primates, such as macaques. But more distantly related primates, including those known as prosimians, have fully functional GLO genes. That means that the common ancestor in which the capacity to make vitamin C was originally lost wasn't human, but a primate—an ancestor that, according to the advocates of intelligent design, we're not supposed to have."[13]

Yet intelligent design and common ancestry are two different issues. Major ID proponents pointed this out before Miller wrote his book.[14–18] Indeed, Lehigh University biochemist and prominent ID advocate Michael J. Behe wrote in 1996 that "the simplest possible design scenario posits a single cell—formed billions of years ago—that already contained all information to produce descendant organisms."[19] As we saw in Chapter 1, intelligent design states that we can infer from evidence in nature that some features of the world, and of living things, are better explained by an intelligent cause than by unguided natural processes. Although some ID proponents (including me) question universal common ancestry on empirical grounds (as do some evolutionary biologists),[20–21] intelligent design is not necessarily inconsistent with common ancestry.

In addition to mischaracterizing ID, Miller went well beyond the published scientific evidence available at the time. For example, as of 2008 (when Miller's book appeared), there were no published data on the gorilla's need for dietary vitamin C.[22] Indeed, the most authoritative review of the vitamin C requirements of non-human primates, published by the U. S. National Academy of Sciences in 2003, did not even mention gorillas.[23] Furthermore, when Miller published his book the sequencing of the gorilla genome had not been completed, and no vitamin C pseudogene had been reported.[24] It wasn't until October 2010 that a sequence was published for a gorilla vitamin C pseudogene.[5] For Miller, apparently, it was conclusion first and evidence later.

Jerry Coyne's Argument

In 2009, University of Chicago geneticist Jerry A. Coyne also argued that the vitamin C pseudogene provides evidence for common ancestry. He began by pointing out that the GLO pseudogene "doesn't work because a single nucleotide in the gene's DNA sequence is missing. And it's exactly the *same* nucleotide missing in other primates. This shows that the mutation that destroyed our ability to make vitamin C was present in the ancestor of all primates, and was passed on to its descendants. The inactivation of GLO in guinea pigs happened independently, since it involves different mutations."[25]

Coyne then argued that this is evidence against creation by design. "If you believe that primates and guinea pigs were specially created," he wrote, "these facts don't make any sense. Why would a creator put a pathway for making vitamin C in all these species, and then inactivate it? Wouldn't it be easier simply to omit the whole pathway from the beginning? Why would the same inactivating mechanism be present in all primates, and a different one in guinea pigs? Why would the sequences of the dead gene exactly mirror the pattern of resemblance predicted from the known ancestry of these species?"[26]

Yet other aspects of the genome do *not* mirror the pattern Coyne predicted. For example, the human Y chromosome (which determines male sexual characteristics) contains about 60 million nucleotide subunits. If humans and chimps were recently descended from a common ancestor, one would expect their Y chromosomes to be very similar. Genome researchers recently reported, however, that the male-specific portions of the human and chimp Y chromosomes "differ radically in sequence structure and gene content."[27] If similarities in the vitamin C pseudogene are evidence for common ancestry, then differences in the Y chromosome are presumably evidence against it.

Furthermore, Coyne's argument—like Miller's—went well beyond the scientific evidence. For example, Coyne claimed that "all primates" not only need vitamin C in their diets, but also have "the same inactivating mechanism"—namely, a single missing nucleotide. Yet prosimians (the lemurs and lorises) are primates that synthesize their own vitamin C, as Miller pointed out. And the need for dietary vitamin C has been established for only nine of the over 260 known species of monkeys.[2,23,28] It is quite possible that some—or even many—monkeys can make their own vitamin C. After scientists reported in 1976 that 34 of the over 800 known species of bats lacked the ability to make their own vitamin C,[4] it was assumed for decades that all bats were alike in this respect—yet scientists recently discovered that some bats (not included in the original study) *can* make their own vitamin C.[5]

So Coyne didn't have the evidence to justify his claim that all primates need vitamin C in their diets, and he was even less justified in claiming that they are all missing the same nucleotide in their *GLO* gene. In fact, the only primates for which *GLO* pseudogene sequences have been published are rhesus macaques, orangutans, chimpanzees, humans, and (more recently) gorillas.[5, 29] Furthermore, the inactivation of the *GLO* gene might have been due to something other than the deletion of a single nucleotide. The same scientists who first detected the missing nucleotide in 1999[29] concluded in 2003 that "it is not possible at present to decide what was the primary change responsible for the functional loss of the gene."[30]

Assumptions Masquerading as Evidence?

In addition to going well beyond the scientific evidence, the vitamin C arguments of Miller and Coyne rely on speculations about the motives of the designer or creator that have no legitimate place in natural science. As we saw in Chapter 10, such speculations are common in Darwin's writing and the literature defending his theory. But the normal practice in science is to test hypotheses against evidence from nature, not speculations based on theological assumptions.

Central to the vitamin C arguments of Miller and Coyne is their assumption that the *GLO* pseudogene is completely nonfunctional. To be sure, there is general agreement that the pseudogene does not produce a functional enzyme, but this does not necessarily mean that it is completely without function. Indeed, as we saw in Chapter 5, there is growing evidence that although pseudogenes don't code for proteins they produce RNAs that function in various aspects of gene regulation.

Miller and Coyne have not provided any evidence to justify their assumption that the *GLO* pseudogene is completely nonfunctional. In fact, they cannot. The strongest statement that could be warranted by the evidence would be that we do not currently know of a function for the vitamin C pseudogene.

The Vitamin C Pseudogene Argument is Circular

IF THE *GLO* pseudogene turns out to serve any function at all, then the sequence similarities in humans and chimps on which Miller and Coyne based their arguments could be due to natural selection rather than common ancestry. In fact, as we saw in Chapter 5, Balakirev and Ayala in 2003 and Khachane and Harrison in 2009 argued that similarities in pseudogenes are presumptive evidence that those pseudogenes are functional.[31-32] Why don't Miller and Coyne argue likewise that the similarities in primate vitamin C pseudogenes suggest functionality rather than common ancestry?

The difference is that the organisms analyzed by Balakirev and Ayala (humans, mice, chickens and fruit flies) and Khachane and Harrison (humans, monkeys, mice, rats, dogs and cows)—unlike humans and chimps—are not thought to share a recent common ancestor. In other words, if organisms *are not* thought to be closely related through common descent, then pseudogene similarities imply function, but if organisms *are* thought to be closely related through common descent, then pseudogene similarities imply that they are closely related through common descent. The second form (used by Miller and Coyne) is a circular argument, because the conclusion is already stated in the premises.

To break the circle, Miller and Coyne would either have to establish the recent common ancestry of humans and chimps on other grounds (but then, why bother invoking the vitamin C pseudogene at all?), or they would first have to establish that the vitamin C pseudogene has no function whatsoever (but this is impossible). So their argument not only fails to refute ID, but it also fails to establish that humans and chimps are descended from a common ancestor.

NOTES

1. THE CONTROVERSY OVER DARWINIAN EVOLUTION

1. Theodosius Dobzhansky, *Genetics and the Origin of Species*, Reprinted 1982. (New York: Columbia University Press, 1937), p. 12.

2. Keith Stewart Thomson, "Natural Selection and Evolution's Smoking Gun," *American Scientist* 85 (1997): 516–518.

3. Alan Linton, "Scant Search for the Maker," *The Times Higher Education Supplement* (April 20, 2001), Book Section, p. 29. Freely accessible (2011) at http://www.timeshighereducation.co.uk/story.asp?storyCode=159282&sectioncode=31

4. Jonathan Wells, *The Politically Incorrect Guide to Darwinism and Intelligent Design* (Washington, DC: Regnery Publishing, 2006), Chapter 5. More information available online (2011) at http://www.discovery.org/a/3699

5. Jonathan Wells, *Icons of Evolution: Science or Myth?* (Washington, DC: Regnery Publishing, 2000). More information available online (2011) at http://www.iconsofevolution.com/

6. Charles Darwin, *The Origin of Species by Means of Natural Selection*, First Edition (London: John Murray, 1859), p. 130. Freely accessible (2011) at http://darwin-online.org.uk/content/frameset?viewtype=side&itemID=F373&pageseq=148

7. Darwin, *The Origin of Species*, p. 282. Freely accessible (2010) at http://darwin-online.org.uk/content/frameset?viewtype=side&itemID=F373&pageseq=300

8. James W. Valentine, Stanley M. Awramik, Philip W. Signor and Peter M. Sadler, "The Biological Explosion at the Precambrian-Cambrian Boundary," *Evolutionary Biology* 25 (1991): 279–356.

9. Jeffrey S. Levinton, "The Big Bang of Animal Evolution," *Scientific American* 267 (November, 1992): 84–91.

10. Jonathan Wells, "Deepening Darwin's Dilemma," *Discovery Institute* (September 16, 2009). Freely accessible (2011) at http://www.discovery.org/a/12471

11. W. Ford Doolittle, "The practice of classification and the theory of evolution, and what the demise of Charles Darwin's tree of life hypothesis means for both of them," *Philosophical Transactions of the Royal Society of London B* 364 (2009): 2221–2228.

12. Carl R. Woese & Nigel Goldenfeld, "How the Microbial World Saved Evolution from the Scylla of Molecular Biology and the Charybdis of the Modern Synthesis," *Microbiology and Molecular Biology Reviews* 73 (2009): 14–21. Freely accessible (2011) at http://mmbr.asm.org/cgi/reprint/73/1/14

13. Wells, *The Politically Incorrect Guide to Darwinism and Intelligent Design*, Chapter 4.

14. Gavin de Beer, *Homology: An Unsolved Problem* (London: Oxford University Press, 1971), pp. 15–16.

15. Wells, *Icons of Evolution: Science or Myth?*, Chapter 4.

16. Charles Darwin, "Letter to Asa Gray, September 10, 1860," in Francis Darwin (editor), *The Life and Letters of Charles Darwin* (London: John Murray, 1887), Vol. II, p. 338. Freely accessible (2011) at http://darwin-online.org.uk/content/frameset?viewtype=side&itemID=F1452.2&pageseq=354

17. Rudolf A. Raff, *The Shape of Life: Genes, Development, and the Evolution of Animal Form* (Chicago: The University of Chicago Press, 1996), pp. 195, 208–209.

18. Jonathan Wells, "Haeckel's Embryos & Evolution: Setting the Record Straight," *The American Biology Teacher* 61 (May, 1999): 345–349. Freely accessible (2011) at http://www.discovery.org/a/3071

19. Wells, *Icons of Evolution: Science or Myth?* Chapter 5.

2. Junk DNA – The Last Icon of Evolution?

1. Horace Freeland Judson, *The Eighth Day of Creation* (New York: Simon and Schuster, 1979), p. 175.

2. Francis Darwin (editor), *The Life and Letters of Charles Darwin* (London: John Murray, 1887), Volume I, p. 309. Freely accessible (2011) at http://darwin-online.org.uk/content/frameset?viewtype=side&itemID=F1452.1&pageseq=327

3. Francis Darwin & A.C. Seward (editors), *More Letters of Charles Darwin* (London: John Murray, 1903), Volume 1, p. 321. Freely accessible (2011) at http://darwin-online.org.uk/content/frameset?viewtype=side&itemID=F1548.1&pageseq=370

4. Francis Darwin (editor), *The Life and Letters of Charles Darwin* (London: John Murray, 1887), Volume II, p. 312. Freely accessible (2011) at http://darwin-online.org.uk/content/frameset?viewtype=side&itemID=F1452.2&pageseq=328

5. William Bateson, *Mendel's Principles of Heredity* (New York: G. P. Putnam's Sons, 1913), p. 329.

6. "Mendel, Mendelism," *The Catholic Encyclopedia*. Freely accessible (2011) at http://www.newadvent.org/cathen/10180b.htm

7. James D. Watson & Francis H. C. Crick, "Molecular Structure of Nucleic Acids: A Structure for Deoxyribose Nucleic Acid," *Nature* 171 (1953): 737–738. Freely accessible (2011) at http://www.annals.org/cgi/reprint/138/7/581.pdf

8. James D. Watson & Francis H. C. Crick, "Genetical Implications of the Structure of Deoxyribonucleic Acid," *Nature* 171 (1953): 964–967.

9. Francis H. C. Crick, "On Protein Synthesis," *The Biological Replication of Macromolecules*, Symposia of the Society for Experi-

mental Biology, Number XII (Cambridge: Cambridge University Press, 1958), pp. 138–163.

10. Judson, *The Eighth Day of Creation*, p. 217.

11. Richard Dawkins, *The Selfish Gene* (New York: Oxford University Press, 1976), pp. 2, 24–25.

12. Susumu Ohno, "So much 'junk' DNA in our genome," *Brookhaven Symposia in Biology* 23 (1972): 366–70. Freely accessible (2011) at http://www.junkdna.com/ohno.html

13. David E. Comings, "The Structure and Function of Chromatin," *Advances in Human Genetics* 3 (1972): 237–431.

14. Dawkins, *The Selfish Gene*, p. 47.

15. W. Ford Doolittle & Carmen Sapienza, "Selfish genes, the phenotype paradigm and genome evolution," *Nature* 284 (1980): 601–603.

16. Leslie E. Orgel & Francis H. C. Crick, "Selfish DNA: the ultimate parasite," *Nature* 284 (1980): 604–607.

17. Thomas Cavalier-Smith, "How selfish is DNA?" *Nature* 285 (1980): 617–618.

18. Gabriel Dover, "Ignorant DNA?" *Nature* 285 (1980): 618–620.

19. Charles B. Thaxton, Walter L. Bradley & Roger L. Olsen, *The Mystery of Life's Origin* (Dallas, TX: Lewis and Stanley, 1984), pp. 210–211.

20. Michael Denton, *Evolution: A Theory in Crisis* (Bethesda, MD: Adler & Adler, 1985), p. 341.

21. Phillip E. Johnson, *Darwin On Trial.* (Washington, DC: Regnery Gateway, 1991), p. 144.

22. Kenneth R. Miller, "Life's Grand Design," *Technology Review* 97 (February–March 1994): 24–32. Freely accessible (2011) at http://www.millerandlevine.com/km/evol/lgd/index.html

23. Richard Dawkins, *A Devil's Chaplain: Reflections on Hope, Lies, Science, and Love* (New York: Mariner Books, 2004), p. 99.

24. Douglas J. Futuyma, *Evolution* (Sunderland, MA: Sinauer Associates, 2005), pp. 48–49, 456, 530.

25. Michael Shermer, *Why Darwin Matters: The Case Against Intelligent Design* (New York: Holt, 2006), pp. 74–75.

26. Francis S. Collins, *The Language of God: A Scientist Presents Evidence for Belief* (New York: Free Press, 2006), pp. 136–137.

27. Philip Kitcher, *Living With Darwin: Evolution, Design, and the Future of Faith* (New York: Oxford, 2007), pp. 57–58, 111.

28. Kenneth R. Miller, *Only a Theory: Evolution and the Battle for America's Soul* (New York: Viking, 2008), pp. 97–98.

29. Jerry A. Coyne, *Why Evolution Is True* (New York: Viking, 2009), pp. 66–67, 81.

30. Richard Dawkins, *The Greatest Show on Earth: The Evidence for Evolution* (New York: Free Press, 2009), pp. 332–333.

31. John C. Avise, *Inside the Human Genome: A Case for Non-Intelligent Design* (Oxford: Oxford University Press, 2010), pp. 82, 115.

32. John C. Avise, "Footprints of nonsentient design inside the human genome," *Proceedings of the National Academy of Sciences USA* 107 Supplement 2 (2010): 8969–8976. Freely accessible (2011) at http://www.pnas.org/content/107/suppl.2/8969.full.pdf+html

3. Most DNA Is Transcribed into RNA

1. Francis H. C. Crick, "On Protein Synthesis," *The Biological Replication of Macromolecules*, Symposia of the Society for Experimental Biology, Number XII (Cambridge: Cambridge University Press, 1958), pp. 138–163.

2. C. Mulder, J. R. Arrand, H. Delius, W. Keller, U. Pettersson, R. J. Roberts & P. A. Sharp, "Cleavage Maps of DNA from Adenovirus Types 2 and 5 by Restriction Endonucleases *EcoRI* and *HpaI*," *Cold Spring Harbor Symposia on Quantitative Biology* 39 (1975): 397–400.

3. Nobel Prize for Physiology or Medicine (1993) awarded to Richard J. Roberts and Phillip A. Sharp for their "discovery of split genes." Press release available online (2011) at http://nobelprize.org/nobel_prizes/medicine/laureates/1993/press.html

4. David M. Glover & David S. Hogness, "A Novel Arrangement of the 18s and 28s Sequences in a Repeating Unit of *Drosophila melanogaster* rDNA," *Cell* 10 (1977): 167–176.

5. Walter Gilbert, "Why genes in pieces?" *Nature* 271 (1978): 501.

6. P. M. B. Walker & Anne McLaren, "Fractionation of mouse deoxyribonucleic acid on hydroxyapatite," *Nature* 208 (1965): 1175–1179.

7. Roy J. Britten & D. E. Kohne, "Repeated Sequences in DNA," *Science* 161 (1968): 529–540.

8. Reviewed in W. G. Flamm, "Highly Repetitive Sequences of DNA in Chromosomes," *International Review of Cytology* 32 (1972): 1–51.

9. Bruce Alberts, Alexander Johnson, Julian Lewis, Martin Raff, Keith Roberts & Peter Walter, *Molecular Biology of the Cell*, Fourth Edition (New York: Garland Science, 2002), p. 203.

10. Joshua Lederberg & Alexa T. McCray, " 'Ome Sweet 'Omics—A Genealogical Treasury of Words," *The Scientist* 15 (2001): 8. Freely accessible (2011) at http://www.lhncbc.nlm.nih.gov/lhc/docs/published/2001/pub2001047.pdf

11. Edmund Pillsbury, "A History of Genome Sequencing," Computational Biology and Bioinformatics, Yale University (1997). Freely accessible (2011) at http://bioinfo.mbb.yale.edu/course/projects/final-4/

12. National Center for Biotechnology Information (GenBank). http://www.ncbi.nlm.nih.gov/genbank/

13. EMBL Nucleotide Sequence Database. http://www.ebi.ac.uk/embl/

14. DNA Data Bank of Japan. http://www.ddbj.nig.ac.jp/

15. "International Consortium Completes Human Genome Project," National Human Genome Research Institute, Bethesda, MD (April 14, 2003). Freely accessible (2011) at http://www.genome.gov/11006929

16. "The ENCODE Project," National Human Genome Research Institute, Bethesda, MD (December 28, 2009). Freely accessible (2011) at http://www.genome.gov/10005107

17. "History," RIKEN Omic Sciences Center, Yokohama, Japan (2009). Freely accessible (2011) at http://www.osc.riken.jp/english/outline/history/

18. FANTOM Consortium, Yokohama, Japan. http://fantom.gsc.riken.jp/4/

19. Fred A. Wright, William J. Lemon, Wei D. Zhao, Russell Sears, Degen Zhuo, Jian-Ping Wang, Hee-Yung Yang, Troy Baer, Don Stredney, Joe Spitzner, Al Stutz, Ralf Krahe & Bo Yuan, "A draft annotation and overview of the human genome," Genome Biology 2:7 (2001). Freely accessible (2011) at http://genomebiology.com/content/pdf/gb-2001-2-7-research0025.pdf

20. Y. Okazaki, M. Furuno, T. Kasukawa, J. Adachi, H. Bono, S. Kondo, I. Nikaido, N. Osato, R. Saito, H. Suzuki, I. Yamanaka, H. Kiyosawa, K. Yagi, Y. Tomaru, Y. Hasegawa, A. Nogami, C. Schönbach, T. Gojobori, R. Baldarelli, D. P. Hill, C. Bult, D. A. Hume, J. Quackenbush, L. M. Schriml, A. Kanapin, H. Matsuda, S. Batalov, K. W. Beisel, J. A. Blake, D. Bradt, V. Brusic, C. Chothia, L. E. Corbani, S. Cousins, E. Dalla, T. A. Dragani, C. F. Fletcher, A. Forrest, K. S. Frazer, T. Gaasterland, M. Gariboldi, C. Gissi, A. Godzik, J. Gough, S. Grimmond, S. Gustincich, N. Hirokawa, I. J. Jackson, E. D. Jarvis, A. Kanai, H. Kawaji, Y. Kawasawa, R. M. Kedzierski, B. L. King, A. Konagaya, I. V. Kurochkin, Y. Lee, B. Lenhard, P. A. Lyons, D. R. Maglott, L. Maltais, L. Marchionni, L. McKenzie, H. Miki, T. Nagashima, K. Numata, T. Okido, W. J. Pavan, G. Pertea, G. Pesole, N. Petrovsky, R. Pillai, J. U. Pontius, D. Qi, S. Ramachandran, T. Ravasi, J. C. Reed, D. J. Reed, J. Reid, B. Z. Ring, M. Ringwald, A. Sandelin, C. Schneider, C. A. M. Semple, M. Setou, K. Shimada, R. Sultana, Y. Takenaka, M. S. Taylor, R. D. Teasdale, M. Tomita, R. Verardo, L. Wagner, C. Wahlestedt, Y. Wang, Y. Watanabe, C. Wells, L. G. Wilming, A. Wynshaw-Boris, M. Yanagisawa, I. Yang, L. Yang, Z. Yuan, M. Zavolan, Y. Zhu, A. Zimmer, P. Carninci, N. Hayatsu, T. Hirozane-Kishikawa, H. Konno, M. Nakamura, N. Sakazume, K. Sato, T. Shiraki, K. Waki, J. Kawai, K. Aizawa, T. Arakawa, S. Fukuda, A. Hara, W. Hashizume, K. Imotani, Y. Ishii, M. Itoh, I. Kagawa, A. Miyazaki, K. Sakai, D. Sasaki, K. Shibata, A. Shinagawa, A. Yasunishi, M. Yoshino, R. Waterston, E. S. Lander, J. Rogers, E. Birney & Y. Hayashizaki, "Analysis of the mouse transcriptome based on functional annotation of 60,770 full-length cDNAs," Nature 420 (2002): 563–573.

21. Philipp Kapranov, Simon E. Cawley, Jorg Drenkow, Stefan Bekiranov, Robert L. Strausberg, Stephen P. A. Fodor & Thomas R. Gingeras, "Large-Scale Transcriptional Activity in Chromosomes 21 and 22," Science 296 (2002): 916–919.

22. P. Carninci, T. Kasukawa, S. Katayama, J. Gough, M. C. Frith, N. Maeda, R. Oyama, T. Ravasi, B. Lenhard, C. Wells, R. Kodzius, K. Shimokawa, V. B. Bajic, S. E. Brenner, S. Batalov, A. R. R. Forrest, M. Zavolan, M. J. Davis, L. G. Wilming, V. Aidinis, J. E. Allen, A. Ambesi-Impiombato, R. Apweiler, R. N. Aturaliya, T. L. Bailey, M. Bansal, L. Baxter, K. W. Beisel, T. Bersano, H. Bono, A. M. Chalk, K. P. Chiu, V. Choudhary, A. Christofels, D. R. Clutterbuck, M. L. Crowe, E.

Dalla, B. P. Dalrymple, B. de Bono, G. Della Gatta, D. di Bernardo, T. Down, P. Engstrom, M. Fagiolini, G. Faulkner, C. F. Fletcher, T. Fukushima, M. Furuno, S. Futaki, M. Gariboldi, P. Georgii-Hemming, T. R. Gingeras, T. Gojobori, R. E. Green, S. Gustincich, M. Harbers, Y. Hayashi, T. K. Hensch, N. Hirokawa, D. Hill, L. Huminiecki, M. Iacono, K. Ikeo, A. Iwama, T. Ishikawa, M. Jakt, A. Kanapin, M. Katoh, Y. Kawasawa, J. Kelso, H. Kitamura, H. Kitano, G. Kollias, S. P. T. Krishnan, A. Kruger, S. K. Kummerfeld, I. V. Kurochkin, L. F. Lareau, D. Lazarevic, L. Lipovich, J. Liu, S. Liuni, S. McWilliam, M. Madan Babu, M. Madera, L. Marchionni, H. Matsuda, S. Matsuzawa, H. Miki, F. Mignone, S. Miyake, K. Morris, S. Mottagui-Tabar, N. Mulder, N. Nakano, H. Nakauchi, P. Ng, R. Nilsson, S. Nishiguchi, S. Nishikawa, F. Nori, O. Ohara, Y. Okazaki, V. Orlando, K. C. Pang, W. J. Pavan, G. Pavesi, G. Pesole, N. Petrovsky, S. Piazza, J. Reed, J. F. Reid, B. Z. Ring, M. Ringwald, B. Rost, Y. Ruan, S. L. Salzberg, A. Sandelin, C. Schneider, C. Schönbach, K. Sekiguchi, C. A. M. Semple, S. Seno, L. Sessa, Y. Sheng, Y. Shibata, H. Shimada, K. Shimada, D. Silva, B. Sinclair, S. Sperling, E. Stupka, K. Sugiura, R. Sultana, Y. Takenaka, K. Taki, K. Tammoja, S. L. Tan, S. Tang, M. S. Taylor, J. Tegner, S. A. Teichmann, H. R. Ueda, E. van Nimwegen, R. Verardo, C. L. Wei, K. Yagi, H. Yamanishi, E. Zabarovsky, S. Zhu, A. Zimmer, W. Hide, C. Bult, S. M. Grimmond, R. D. Teasdale, E. T. Liu, V. Brusic, J. Quackenbush, C. Wahlestedt, J. S. Mattick, D. A. Hume, C. Kai, D. Sasaki, Y. Tomaru, S. Fukuda, M. Kanamori-Katayama, M. Suzuki, J. Aoki, T. Arakawa, J. Iida, K. Imamura, M. Itoh, T. Kato, H. Kawaji, N. Kawagashira, T. Kawashima, M. Kojima, S. Kondo, H. Konno, K. Nakano, N. Ninomiya, T. Nishio, M. Okada, C. Plessy, K. Shibata, T. Shiraki, S. Suzuki, M. Tagami, K. Waki, A. Watahiki, Y. Okamura-Oho, H. Suzuki, J. Kawai & Y. Hayashizaki, "The Transcriptional Landscape of the Mammalian Genome," *Science* 309 (2005): 1559–1563. Available online with registration (2011) at http://www.sciencemag.org/cgi/content/full/309/5740/1559

23. Michael Pheasant & John S. Mattick, "Raising the estimate of functional human sequences," *Genome Research* 17 (2007): 1245–1253. Freely accessible (2011) at http://genome.cshlp.org/content/17/9/1245.full.pdf+html

24. Ewan Birney, John A. Stamatoyannopoulos, Anindya Dutta, Roderic Guigó, Thomas R. Gingeras, Elliott H. Margulies, Zhiping Weng, Michael Snyder, Emmanouil T. Dermitzakis, Robert E. Thurman, Michael S. Kuehn, Christopher M. Taylor, Shane Neph, Christoph M. Koch, Saurabh Asthana, Ankit Malhotra, Ivan Adzhubei, Jason A. Greenbaum, Robert M. Andrews, Paul Flicek, Patrick J. Boyle, Hua Cao, Nigel P. Carter, Gayle K. Clelland, Sean Davis, Nathan Day, Pawandeep Dhami, Shane C. Dillon, Michael O. Dorschner, Heike Fiegler, Paul G. Giresi, Jeff Goldy, Michael Hawrylycz, Andrew Haydock, Richard Humbert, Keith D. James, Brett E. Johnson, Ericka M. Johnson, Tristan T. Frum, Elizabeth R. Rosenzweig, Neerja Karnani, Kirsten Lee, Gregory C. Lefebvre, Patrick A. Navas, Fidencio Neri, Stephen C. J. Parker, Peter J. Sabo, Richard Sandstrom, Anthony Shafer, David Vetrie, Molly Weaver, Sarah Wilcox, Man Yu, Francis S. Collins, Job Dekker, Jason D. Lieb, Thomas D. Tullius, Gregory E. Crawford, Shamil Sunayev, William S. Noble, Ian Dunham, France Denoeud, Alexandre Reymond, Philipp Kapranov, Joel Rozowsky, Deyou Zheng, Robert Castelo, Adam Frankish, Jennifer Harrow, Srinka Ghosh, Albin Sandelin, Ivo L. Hofacker, Robert Baertsch, Damian Keefe, Sujit Dike, Jill Cheng, Heather A. Hirsch, Edward A. Sekinger, Julien Lagarde, Josep F. Abril, Atif Shahab, Christoph Flamm, Claudia Fried, Jörg Hack-

ermüller, Jana Hertel, Manja Lindemeyer, Kristin Missal, Andrea Tanzer, Stefan Washietl, Jan Korbel, Olof Emanuelsson, Jakob S. Pedersen, Nancy Holroyd, Ruth Taylor, David Swarbreck, Nicholas Matthews, Mark C. Dickson, Daryl J. Thomas, Matthew T. Weirauch, James Gilbert, Jorg Drenkow, Ian Bell, XiaoDong Zhao, K. G. Srinivasan, Wing-Kin Sung, Hong Sain Ooi, Kuo Ping Chiu, Sylvain Foissac, Tyler Alioto, Michael Brent, Lior Pachter, Michael L. Tress, Alfonso Valencia, Siew Woh Choo, Chiou Yu Choo, Catherine Ucla, Caroline Manzano, Carine Wyss, Evelyn Cheung, Taane G. Clark, James B. Brown, Madhavan Ganesh, Sandeep Patel, Hari Tammana, Jacqueline Chrast, Charlotte N. Henrichsen, Chikatoshi Kai, Jun Kawai, Ugrappa Nagalakshmi, Jiaqian Wu, Zheng Lian, Jin Lian, Peter Newburger, Xueqing Zhang, Peter Bickel, John S. Mattick, Piero Carninci,Yoshihide Hayashizaki, Sherman Weissman, Tim Hubbard, Richard M. Myers, Jane Rogers, Peter F. Stadler, Todd M. Lowe, Chia-Lin Wei, Yijun Ruan, Kevin Struhl, Mark Gerstein, Stylianos E. Antonarakis, Yutao Fu, Eric D. Green, Ulaf Karaöz, William S. Noble, Alexandre Reymond, Adam Siepel, James Taylor, Thomas D. Tullius, Laura A. Liefer, Kris A. Wetterstrand, Peter J. Good, Elise A. Feingold, Mark S. Guyer, Gregory M. Cooper, George Asimenos, Daryl J. Thomas, Colin N. Dewey, Minmei Hou, Sergey Nikolaev, Juan I. Montoya-Burgos, Ari Löytynoja, Simon Whelan, Fabio Pardi, Tim Massingham, Haiyan Huang, Nancy R. Zhang, Ian Holmes, James C. Mullikin, Abel Ureta-Vidal, Benedict Paten, Michael Seringhaus, Deanna Church, Kate Rosenbloom, W. James Kent, Serafim Batzoglou, Nick Goldman, Ross C. Hardison, David Haussler, Webb Miller, Lior Pachter, Arend Sidow, Gerard G. Bouffard, Xiaobin Guan, Nancy F. Hansen, Jacquelyn R. Idol, Valerie V.B. Maduro, Baishali Maskeri, Jennifer C. McDowell, Morgan Park, Pamela J. Thomas, Alice C. Young, Robert W. Blakesley, Donna M. Muzny, Erica Sodergren, David A. Wheeler, Kim C. Worley, Huaiyang Jiang, George M. Weinstock, Richard A. Gibbs, Tina Graves, Robert Fulton, Elaine R. Mardis, Richard K. Wilson, Michele Clamp, James Cuff, Sante Gnerre, David B. Jaffe, Jean L. Chang, Kerstin Lindblad-Toh, Eric S. Lander, Maxim Koriabine, Mikhail Nefedov, Kazutoyo Osoegawa, Yuko Yoshinaga, Baoli Zhu, Pieter J. de Jong, Nathan D. Trinklein, Zhengdong D. Zhang, Leah Barrera, Rhona Stuart, David C. King, Adam Ameur, Stefan Enroth, Mark C. Bieda, Chia-Lin Wei, Jonghwan Kim, Akshay A. Bhinge, Paul G. Giresi, Nan Jiang, Jun Liu, Fei Yao, Wing-Kin Sung, Kuo Ping Chiu, Vinsensius B. Vega, Charlie W.H. Lee, Patrick Ng, Atif Shahab, Edward A. Sekinger, Annie Yang, Zarmik Moqtaderi, Zhou Zhu, Xiaoqin Xu, Sharon Squazzo, Matthew J. Oberley, David Inman, Michael A. Singer, Todd A. Richmond, Kyle J. Munn, Alvaro Rada-Iglesias, Ola Wallerman, Jan Komorowski, Gayle K. Clelland, Robert M. Andrews, Joanna C. Fowler, Phillippe Couttet, Keith D. James, Gregory C. Lefebvre, Alexander W. Bruce, Oliver M. Dovey, Peter D. Ellis, Pawandeep Dhami, Cordelia F. Langford, Nigel P. Carter, David Vetrie, David A. Nix, Ian Bell, Ghia Euskirchen, Stephen Hartman, Jiaqian Wu, Alexander E. Urban, Peter Kraus, Sara Van Calcar, Nate Heintzman, Tae Hoon Kim, Kun Wang, Chunxu Qu, Gary Hon, Rosa Luna, Christopher K. Glass, M. Geoff Rosenfeld, Shelley Force Aldred, Sara J. Cooper, Anason Halees, Jane M. Lin, Hennady P. Shulha, Xiaoling Zhang, Mousheng Xu, Jaafar N. S. Haidar, Yong Yu, Sherman Weissman, Yijun Ruan, Jason D. Lieb, Vishwanath R. Iyer, Roland D. Green, Claes Wadelius, Ian Dunham, Peggy J. Farnham, Bing Ren, Rachel A. Harte, Angie S. Hinrichs, Heather Trumbower, Hiram Clawson, Jennifer Hillman-Jackson, Ann S. Zweig, Kayla Smith, Archana Thakkapallayil, Galt Barber, Robert M. Kuhn, Donna Karolchik, W. James Kent, Lluis Armengol, Christine P. Bird, Taane G. Clark, Paul I. W. de Bakker, Andrew D. Kern, Nuria Lopez-Bigas, Joel D. Martin, Barbara E. Stranger, Abigail Woodroffe, Serafim Batzoglou, Eugene Davydov, Antigone Dimas, Eduardo Eyras, Ingileif B. Hallgrímsdóttir, Julian Huppert, Heather Trumbower, Michael C. Zody, James C. Mullikin, Gonçalo R. Abecasis & Xavier Estivill, "Identification and analysis of functional elements in 1% of the human genome by the ENCODE pilot project," *Nature* 447 (2007): 799–816. Freely accessible (2011) at http://www.ncbi.nlm.nih. gov/pmc/articles/PMC2212820/pdf/ nihms27513.pdf

25. Naoki Osato, Hitomi Yamada, Kouji Satoh, Hisako Ooka, Makoto Yamamoto, Kohji Suzuki, Jun Kawai, Piero Carninci, Yasuhiro Ohtomo, Kazuo Murakami, Kenichi Matsubara, Shoshi Kikuchi & Yoshihide Hayashizaki, "Antisense transcripts with rice full-length cDNAs," *Genome Biology* 5:1 (2003): R5. Freely accessible (2011) at http://genomebiology. com/content/pdf/gb-2003-5-1-r5.pdf

26. S. Katayama, Y. Tomaru, T. Kasukawa, K. Waki, M. Nakanishi, M. Nakamura, H. Nishida, C. C. Yap, M. Suzuki, J. Kawai, H. Suzuki, P. Carninci, Y. Hayashizaki, C. Wells, M. Frith, T. Ravasi, K. C. Pang, J. Hallinan, J. Mattick, D. A. Hume, L. Lipovich, S. Batalov, P. G. Engström, Y. Mizuno, M. A. Faghihi, A. Sandelin, A. M. Chalk, S. Mottagui-Tabar, Z. Liang, B. Lenhard & C. Wahlestedt, "Antisense Transcription in the Mammalian Transcriptome," *Science* 309 (2005): 1564–1566.

27. Pär G. Engström, Harukazu Suzuki, Noriko Ninomiya, Altuna Akalin, Luca Sessa, Giovanni Lavorgna, Alessandro Brozzi, Lucilla Luzi, Sin Lam Tan, Liang Yang, Galih Kunarso, Edwin Lian-Chong Ng, Serge Batalov, Claes Wahlestedt, Chikatoshi Kai, Jun Kawai, Piero Carninci, Yoshihide Hayashizaki, Christine Wells, Vladimir B. Bajic, Valerio Orlando, James F. Reid, Boris Lenhard & Leonard Lipovich, "Complex Loci in Human and Mouse Genomes," *PLoS Genetics* 2:4 (2006): e47. Freely accessible (2011) at http://www.plosgenetics.org/article/ info%3Adoi%2F10.1371%2Fjournal. pgen.0020047

28. Yiping He, Bert Vogelstein, Victor E. Velculescu, Nickolas Papadopoulos & Kenneth W. Kinzler, "The Antisense Transcriptomes of Human Cells," *Science* 322 (2008): 1855–1857.

29. Kevin V. Morris, Sharon Santoso, Anne-Marie Turner, Chiara Pastori, Peter G. Hawkins, "Bidirectional Transcription Directs Both Transcriptional Gene Activation and Suppression in Human Cells," *PLoS Genetics* 4:11 (2008): e1000258. Freely accessible (2011) at http://www.plosgenetics.org/ article/info:doi%2F10.1371%2Fjournal. pgen.1000258

30. Stefano Gustincich, Albin Sandelin, Charles Plessy, Shintaro Katayama, Roberto Simone, Dejan Lazarevic, Yoshihide Hayashizaki & Piero Carninci, "The complexity of the mammalian transcriptome," *Journal of Physiology* 575:2 (2006): 321–332. Freely accessible (2011) at http:// jp.physoc.org/content/575/2/321.full. pdf+html

31. Philipp Kapranov, Aarron T. Willingham & Thomas R. Gingeras, "Genome-wide transcription and the implications for genomic organization," *Nature Reviews Genetics* 8 (2007): 413–423.

32. Piero Carninci, "Constructing the landscape of the mammalian transcriptome," *Journal of Experimental Biology* 210 (2007): 1497–1506. Freely accessible (2011) at http://jeb.biologists.org/cgi/ reprint/210/9/1497

33. Jia Qian Wu, Jiang Du, Joel Rozowsky, Zhengdong Zhang, Alexander E. Urban,

Ghia Euskirchen, ShermanWeissman, Mark Gerstein & Michael Snyder, "Systematic analysis of transcribed loci in ENCODE regions using RACE sequencing reveals extensive transcription in the human genome," *Genome Biology* 9:1 (2008): R3. Freely accessible (2011) at http://genomebiology.com/content/pdf/gb-2008-9-1-r3.pdf

34. Gill Bejerano, Michael Pheasant, Igor Makunin, Stuart Stephen, W. James Kent, John S. Mattick & David Haussler, "Ultraconserved Elements in the Human Genome," *Science* 304 (2004): 1321–1325.

35. Albin Sandelin, Peter Bailey, Sara Bruce, Pär G. Engström, Joanna M Klos, Wyeth W. Wasserman, Johan Ericson & Boris Lenhard, "Arrays of ultraconserved noncoding regions span the loci of key developmental genes in vertebrate genomes," *BMC Genomics* 5 (2004): 99. Freely accessible (2011) at http://www.biomedcentral.com/1471-2164/5/99

36. Adam Woolfe, Martin Goodson, Debbie K. Goode, Phil Snell, Gayle K. McEwen, Tanya Vavouri, Sarah F. Smith, Phil North, Heather Callaway, Krys Kelly, Klaudia Walter, Irina Abnizova, Walter Gilks, Yvonne J. K. Edwards, Julie E. Cooke & Greg Elgar, "Highly Conserved Non-coding Sequences Are Associated with Vertebrate Development," *PLoS Biology* 3:1 (2005): e7. Freely accessible (2011) at http://www.plosbiology.org/article/info%3Adoi%2F10.1371%2Fjournal.pbio.0030007

37. Adam Siepel, Gill Bejerano, Jakob S. Pedersen, Angie S. Hinrichs, Minmei Hou, Kate Rosenbloom, Hiram Clawson, John Spieth, LaDeana W. Hillier, Stephen Richards, George M. Weinstock, Richard K. Wilson, Richard A. Gibbs, W. James Kent, Webb Miller & David Haussler, "Evolutionarily conserved elements in vertebrate, insect, worm, and yeast genomes," *Genome Research* 15 (2005): 1034–1050. Freely accessible (2011) at http://genome.cshlp.org/content/15/8/1034.full.pdf+html

38. Gil Bejerano, "Ultraconservation and the Human Genome Regulatory Landscape," Lecture at Stanford University (April 15, 2009). Freely accessible (2011) at http://video.google.com/videoplay?docid=8213646681956800413#

39. John A. Bernat, Gregory E. Crawford, Aleksey Y. Ogurtsov, Francis S. Collins, David Ginsburg & Alexey S. Kondrashov, "Distant conserved sequences flanking endothelial-specific promoters contain tissue-specific DNase-hypersensitive sites and over-represented motifs," *Human Molecular Genetics* 15 (2006): 2098–2105. Freely accessible (2011) at http://hmg.oxfordjournals.org/cgi/reprint/15/13/2098

40. Tanya Vavouri, Klaudia Walter, Walter R Gilks, Ben Lehner and Greg Elgar, "Parallel evolution of conserved non-coding elements that target a common set of developmental regulatory genes from worms to humans," *Genome Biology* 8:2 (2007): R15. Freely accessible (2011) at http://genomebiology.com/2007/8/2/R15

41. Jasmina Ponjavic, Chris P. Ponting & Gerton Lunter, "Functionality or transcriptional noise? Evidence for selection within long noncoding RNAs," *Genome Research* 17 (2007): 556–565. Freely accessible (2011) at http://genome.cshlp.org/content/17/5/556.full.pdf+html

42. Mitchell Guttman, Ido Amit, Manuel Garber, Courtney French, Michael F. Lin, David Feldser, Maite Huarte, Or Zuk, Bryce W. Carey, John P. Cassady, Moran N. Cabili, Rudolf Jaenisch, Tarjei S. Mikkelsen, Tyler Jacks, Nir Hacohen, Bradley E. Bernstein, Manolis Kellis, Aviv Regev, John L. Rinn & Eric S. Lander, "Chromatin signature reveals over a thousand highly conserved large non-coding RNAs in mammals," *Nature* 458 (2009): 223–227. Freely accessible (2011) at http://www.ncbi.nlm.nih.gov/pmc/articles/PMC2754849/?tool=pubmed

43. Maciej Szymanski, Miroslawa Z. Barciszewska, Marek Zywicki & Jan Barciszewski, "Noncoding RNA transcripts," *Journal of Applied Genetics* 44 (2003): 1–19. Freely accessible (2011) at http://jag.igr.poznan.pl/2003-Volume-44/1/pdf/2003_Volume_44_1-1-19.pdf

44. John S. Mattick & Igor V. Makunin, "Non-coding RNA," *Human Molecular Genetics* 15 (2006): R17-R29. Freely accessible (2011) at http://hmg.oxfordjournals.org/cgi/reprint/15/suppl_1/R17

45. Luis M. Mendes Soares & Juan Valcárcel, "The expanding transcriptome: the genome as the 'Book of Sand,'" *EMBO Journal* 25 (2006): 923–931. Available online with registration (2011) at http://www.nature.com/emboj/journal/v25/n5/full/7601023a.html

46. John L. Rinn, Michael Kertesz, Jordon K. Wang, Sharon L. Squazzo, Xiao Xu, Samantha A. Brugmann, Henry Goodnough, Jill A. Helms, Peggy J. Farnham, Eran Segal & Howard Y. Chang, "Functional Demarcation of Active and Silent Chromatin Domains in Human HOX Loci by Non-Coding RNAs," *Cell* 129 (2007): 1311–1323. Freely accessible (2011) at http://www.ncbi.nlm.nih.gov/pmc/articles/PMC2084369/?tool=pubmed

47. Gennadi V. Glinsky, "Phenotype-defining functions of multiple non-coding RNA pathways," *Cell Cycle* 7 (2008): 1630–1639. Freely accessible (2011) at http://www.landesbioscience.com/journals/cc/article/5976/

48. Eugene V. Makeyev & Tom Maniatis, "Multilevel Regulation of Gene Expression by MicroRNAs," *Science* 319 (2008): 1789–1790

49. Paulo P. Amaral, Marcel E. Dinger, Tim R. Mercer & John S. Mattick, "The Eukaryotic Genome as an RNA Machine," *Science* 319 (2008): 1787–1789.

50. Tim R. Mercer, Marcel E. Dinger, Susan M. Sunkin, Mark F. Mehler & John S. Mattick, "Specific expression of long noncoding RNAs in the mouse brain," *Proceedings of the National Academy of Sciences USA* 105 (2008): 716–721. Freely accessible (2011) at http://www.pnas.org/content/105/2/716.full.pdf+html

51. Johannes H. Urban & Jörg Vogel, "Two Seemingly Homologous Noncoding RNAs Act Hierarchically to Activate glmS mRNA Translation," *PLoS Biology* 6:3 (2008): e64. Freely accessible (2011) at http://www.plosbiology.org/article/info%3Adoi%2F10.1371%2Fjournal.pbio.0060064

52. Piero Carninci, Jun Yasuda & Yoshihide Hayashizaki, "Multifaceted mammalian transcriptome," *Current Opinion in Cell Biology* 20 (2008): 274–280.

53. Archa H. Fox, Yun Wah Lam, Anthony K. L. Leung, Carol E. Lyon, Jens Andersen, Matthias Mann & Angus I. Lamond, "Paraspeckles: a novel nuclear domain," *Current Biology* 12 (2002): 13–25. Freely accessible (2011) at http://www.cell.com/current-biology/retrieve/pii/S0960982201006327

54. Charles S. Bond & Archa H. Fox, "Paraspeckles: nuclear bodies built on long noncoding RNA," *Journal of Cell Biology* 186 (2009): 637–644. Freely accessible (2011) at http://jcb.rupress.org/content/186/5/637.full.pdf+html

55. Archa H. Fox & Angus I. Lamond, "Paraspeckles," *Cold Spring Harbor Perspectives in Biology* 2 (2010): a000687. Freely accessible (2011) at http://cshperspectives.cshlp.org/content/2/7/a000687.full.pdf+html

56. Christine M. Clemson, John N. Hutchinson, Sergio A. Sara, Alexander W. Ensminger, Archa H. Fox, Andrew Chess & Jeanne B. Lawrence, "An architectural role for a nuclear noncoding RNA: NEAT1 RNA is essential for the structure of paraspeckles," *Molecular Cell* 33 (2009): 717–726. Freely accessible (2011) at http://www.ncbi.nlm.nih.gov/

pmc/articles/PMC2696186/pdf/ni-hms106615.pdf

57. Yasnory T. F. Sasaki, Takashi Ideue, Miho Sano, Toutai Mituyama & Tetsuro Hirose, "MENε/β noncoding RNAs are essential for structural integrity of nuclear paraspeckles," *Proceedings of the National Academy of Sciences USA* 106 (2009): 2525–2530. Freely accessible (2011) at http://www.pnas.org/content/106/8/2525.full.pdf+html

58. Yasnory T. F. Sasaki & Tetsuro Hirose, "How to build a paraspeckle," *Genome Biology* 10 (2009): 227. Freely accessible (2011) at http://genomebiology.com/content/pdf/gb-2009-10-7-227.pdf

59. Sylvie Souquere, Guillaume Beauclair, Francis Harper, Archa Fox & Gérard Pierron, "Highly-ordered spatial organization of the structural long noncoding NEAT1 RNAs within paraspeckle nuclear bodies," *Molecular Biology of the Cell* (September 2010). Freely accessible (2011) at http://www.molbiolcell.org/cgi/reprint/E10-08-0690v1

60. Marcel E. Dinger, Paulo P. Amaral, Timothy R. Mercer & John S. Mattick, "Pervasive transcription of the eukaryotic genome: functional indices and conceptual implications," *Briefings in Functional Genomics and Proteomics* 8 (2009): 407–423.

61. Jeremy E. Wilusz, Hongjae Sunwoo & David L. Spector, "Long noncoding RNAs: functional surprises from the RNA world," *Genes & Development* 23 (2009): 1494–1504. Freely accessible (2011) at http://genesdev.cshlp.org/content/23/13/1494.full.pdf+html

62. Jeannie T. Lee, "Lessons from X-chromosome inactivation: long ncRNA as guides and tethers to the epigenome," *Genes & Development* 23 (2009): 1831–1842. Freely accessible (2011) at http://genesdev.cshlp.org/content/23/16/1831.full.pdf+html

4. INTRONS AND THE SPLICING CODE

1. Stuart E. Leff, Michael G. Rosenfeld & Ronald M. Evans, "Complex transcriptional units: diversity in gene expression by alternative RNA processing," *Annual Review of Biochemistry* 55 (1986): 1091–1117.

2. Richard A. Padgett, Paula J. Grabowski, Maria M. Konarska, Sharon Seiler & Phillip A. Sharp, "Splicing of messenger RNA precursors," *Annual Review of Biochemistry* 55 (1986): 1119–1150.

3. Tom Maniatis & Bosiljka Tasic, "Alternative pre-mRNA splicing and proteome expansion in metazoans," *Nature* 418 (2002): 236–243.

4. Qun Pan, Ofer Shai, Leo J. Lee, Brendan J. Frey & Benjamin J. Blencowe, "Deep surveying of alternative splicing complexity in the human transcriptome by high-throughput sequencing," *Nature Genetics* 40 (2008): 1413–1415.

5. Eric T. Wang, Rickard Sandberg, Shujun Luo, Irina Khrebtukova, Lu Zhang, Christine Mayr, Stephen F. Kingsmore, Gary P. Schroth & Christopher B. Burge, "Alternative isoform regulation in human tissue transcriptomes," *Nature* 456 (2008): 470–476. Freely accessible (2011) at http://www.ncbi.nlm.nih.gov/pmc/articles/PMC2593745/pdf/nihms-72491.pdf

6. Marc Sultan, Marcel H. Schulz, Hugues Richard, Alon Magen, Andreas Klingenhoff, Matthias Scherf, Martin Seifert, Tatjana Borodina, Aleksey Soldatov, Dmitri Parkhomchuk, Dominic Schmidt, Sean O'Keeffe, Stefan Haas, Martin Vingron, Hans Lehrach & Marie-Laure Yaspo, "A Global View of Gene Activity and Alternative Splicing by Deep Sequencing of the Human Transcriptome," *Science* 321 (2008): 956–960. Available online with registration (2011) at http://www.sciencemag.org/content/321/5891/956.short

7. Timothy W. Nilsen & Brenton R. Graveley, "Expansion of the eukaryotic proteome

by alternative splicing," *Nature* 463 (2010): 457–463.

8. Kevin P. Rosenblatt, Zhong-Ping Sun, Stefan Heller & A. J. Hudspeth, "Distribution of Ca²⁺-activated K⁺ channel isoforms along the tonotopic gradient of the chicken's cochlea," *Neuron* 19 (1997): 1061–1075.

9. Dhasakumar S. Navaratnam, Thomas J. Bell, Tu Dinh Tu, Erik L. Cohen & J. Carl Oberholtzer, "Differential distribution of Ca²⁺-activated K⁺ channel splice variants among hair cells along the tonotopic axis of the chick cochlea," *Neuron* 19 (1997): 1077–1085.

10. Dietmar Schmucker, James C. Clemens, Huidy Shu, Carolyn A. Worby, Jian Xiao, Marco Muda, Jack E. Dixon & S. Lawrence Zipursky, "*Drosophila* Dscam is an axon guidance receptor exhibiting extraordinary molecular diversity," *Cell* 101 (2000): 671–684.

11. Kerry Kornfeld, Robert B. Saint, Philip A. Beachy, Peter J. Harte, Debra A. Peattie & David S. Hogness, "Structure and expression of a family of *Ultrabithorax* mRNAs generated by alternative splicing and polyadenylation in *Drosophila*," *Genes & Development* 3 (1989): 243–258. Freely accessible (2011) at http://genesdev.cshlp.org/content/3/2/243.long

12. K. Moriarty, K. H. Kim and J. R. Bender, "Minireview: Estrogen Receptor-Mediated Rapid Signaling," *Endocrinology* 147 (2006): 5557–5563. Freely accessible (2011) at http://endo.endojournals.org/cgi/reprint/147/12/5557

13. Benjamin J. Blencowe, "Alternative splicing: new insights from global analyses," *Cell* 126 (2006): 37–47.

14. Alison Jane Tyson-Capper, "Alternative splicing: an important mechanism for myometrial gene regulation that can be manipulated to target specific genes associated with preterm labour," *BMC Pregnancy Childbirth* 7 Supplement 1 (2007): S13. Freely accessible (2011) at http://www. biomedcentral.com/content/pdf/1471-2393-7-S1-S13.pdf

15. Stefan Hoppler & Claire Louise Kavanagh, "Wnt signalling: variety at the core," *Journal of Cell Science* 120 (2007): 385–93. Freely accessible (2011) at http://jcs.biologists.org/cgi/reprint/120/3/385

16. Antonino Belfiore, Francesco Frasca, Giuseppe Pandini, Laura Sciacca & Riccardo Vigneri, "Insulin Receptor Isoforms and Insulin Receptor/Insulin-like Growth Factor Receptor Hybrids in Physiology and Disease," *Endocrine Reviews* 30 (2009): 586–623. Freely accessible (2011) at http://edrv.endojournals.org/cgi/reprint/30/6/586

17. Ludmila Prokunina-Olsson, Cullan Welch, Ola Hansson, Neeta Adhikari, Laura J. Scott, Nicolle Usher, Maurine Tong, Andrew Sprau, Amy Swift, Lori L. Bonnycastle, Michael R. Erdos, Zhi He, Richa Saxena, Brennan Harmon, Olga Kotova, Eric P. Hoffman, David Altshuler, Leif Groop, Michael Boehnke, Francis S. Collins & Jennifer L. Hall, "Tissue-specific alternative splicing of TCF7L2," *Human Molecular Genetics* 18 (2009): 3795–3804. Freely accessible (2011) at http://hmg.oxfordjournals.org/cgi/reprint/18/20/3795

18. Chiharu Sogawa, Chieko Mitsuhata, Kei Kumagai-Morioka, Norio Sogawa, Kazumi Ohyama, Katsuya Morita, Katsuyuki Kozai, Toshihiro Dohi & Shigeo Kitayama, "Expression and Function of Variants of Human Catecholamine Transporters Lacking the Fifth Transmembrane Region Encoded by Exon 6," *PLoS One* 5:8 (2010): e11945. Freely accessible (2011) at http://www.plosone.org/article/info%3Adoi%2F10.1371%2Fjournal.pone.0011945

19. Anna Kuta, Wenhan Deng, Ali Morsi El-Kadi, Gareth T. Banks, Majid Hafezparast, K. Kevin Pfister & Elizabeth M. C. Fisher, "Mouse Cytoplasmic Dynein Intermediate Chains: Identification of New

Isoforms, Alternative Splicing and Tissue Distribution of Transcripts," *PLoS One* 5:7 (2010): e11682. Freely accessible (2011) at http://www.plosone.org/article/info%3Adoi%2F10.1371%2Fjournal.pone.0011682

20. Ahmet Ucar, Vida Vafaizadeh, Hubertus Jarry, Jan Fiedler, Petra A. B. Klemmt, Thomas Thum, Bernd Groner & Kamal Chowdhury, "miR-212 and miR-132 are required for epithelial stromal interactions necessary for mouse mammary gland development," *Nature Genetics* 42 (2010): 1101–1108.

21. Tim R. Mercer, Marcel E. Dinger, Cameron P. Bracken, Gabriel Kolle, Jan M. Szubert, Darren J. Korbie, Marjan E. Askarian-Amiri, Brooke B. Gardiner, Gregory J. Goodall, Sean M. Grimmond & John S. Mattick, "Regulated post-transcriptional RNA cleavage diversifies the eukaryotic transcriptome," *Genome Research* 20 (2010): 1639–1650.

22. Vidisha Tripathi, Jonathan D. Ellis, Zhen Shen, David Y. Song, Qun Pan, Andrew T. Watt, Susan M. Freier, C. Frank Bennett, Alok Sharma, Paula A. Bubulya, Benjamin J. Blencowe, Supriya G. Prasanth & Kannanganattu V. Prasanth, "The nuclear-retained noncoding RNA MALAT1 regulates alternative splicing by modulating SR splicing factor phosphorylation," *Molecular Cell* 39 (2010): 925–938.

23. Rotem Sorek & Gil Ast, "Intronic Sequences Flanking Alternatively Spliced Exons Are Conserved between Human and Mouse," *Genome Research* 13 (2003): 1631–1637. Freely accessible (2011) at http://genome.cshlp.org/content/13/7/1631.full.pdf+html

24. Simon Minovitsky, Sherry L. Gee, Shiruyeh Schokrpur, Inna Dubchak & John G. Conboy, "The splicing regulatory element, UGCAUG, is phylogenetically and spatially conserved in introns that flank tissue-specific alternative exons," *Nucleic Acids Research* 33 (2005): 714–724.

Freely accessible (2011) at http://nar.oxfordjournals.org/content/33/2/714.full.pdf+html

25. Charles W. Sugnet, Karpagam Srinivasan, Tyson A. Clark, Georgeann O'Brien, Melissa S. Cline, Hui Wang, Alan Williams, David Kulp, John E. Blume, David Haussler & Manuel Ares Jr., "Unusual Intron Conservation near Tissue-regulated Exons Found by Splicing Microarrays," *PLoS Computational Biology* 2:1 (2006): e4. Freely accessible (2011) at http://www.ploscompbiol.org/article/info%3Adoi%2F10.1371%2Fjournal.pcbi.0020004

26. Andrea N. Ladd and Thomas A. Cooper, "Finding signals that regulate alternative splicing in the post-genomic era," *Genome Biology* 3:11 (2002): reviews0008. Freely accessible (2011) at http://genomebiology.com/content/pdf/gb-2002-3-11-reviews0008.pdf

27. Jingyi Hui, Lee-Hsueh Hung, Monika Heiner, Silke Schreiner, Norma Neumüller, Gregor Reither, Stefan A Haas & Albrecht Bindereif, "Intronic CA-repeat and CA-rich elements: a new class of regulators of mammalian alternative splicing," *EMBO Journal* 24 (2005): 1988–1998. Freely accessible (2011) at http://www.nature.com/emboj/journal/v24/n11/pdf/7600677a.pdf

28. Helder I. Nakaya, Paulo P. Amaral, Rodrigo Louro, André Lopes, Angela A. Fachel, Yuri B. Moreira, Tarik A. El-Jundi, Aline M. da Silva, Eduardo M. Reis & Sergio Verjovski-Almeida, "Genome mapping and expression analyses of human intronic noncoding RNAs reveal tissue-specific patterns and enrichment in genes related to regulation of transcription," *Genome Biology* 8:3 (2007): R43. Freely accessible (2011) at http://www.ncbi.nlm.nih.gov/pmc/articles/PMC1868932/pdf/gb-2007-8-3-r43.pdf

29. Michelle L. Hastings, Catherine M. Wilson & Stephen H. Munroe, "A purine-

rich intronic element enhances alternative splicing of thyroid hormone receptor mRNA," *RNA* 7 (2001): 859–874. Freely accessible (2011) at http://www.ncbi.nlm.nih.gov/pmc/articles/PMC1370135/pdf/11421362.pdf

30. Shingo Nakahata & Sachiyo Kawamoto, "Tissue-dependent isoforms of mammalian Fox-1 homologs are associated with tissue-specific splicing activities," *Nucleic Acids Research* 33 (2005): 2078–2089. Freely accessible (2011) at http://www.ncbi.nlm.nih.gov/pmc/articles/PMC1075922/pdf/gki338.pdf

31. Eric J. Wagner, Andrew P. Baraniak, October M. Sessions, David Mauger, Eric Moskowitz & Mariano A. Garcia-Blanco, "Characterization of the Intronic Splicing Silencers Flanking FGFR2 Exon IIIb," *Journal of Biological Chemistry* 280 (2005): 14017–14027. Freely accessible (2011) at http://www.jbc.org/content/280/14/14017.full.pdf+html

32. Roberto Marcucci, Francisco E. Baralle & Maurizio Romano, "Complex splicing control of the human Thrombopoietin gene by intronic G runs," *Nucleic Acids Research* 35 (2007): 132–142. Freely accessible (2011) at http://www.ncbi.nlm.nih.gov/pmc/articles/PMC1802585/pdf/gkl965.pdf

33. Zefeng Wang & Christopher B. Burge, "Splicing regulation: from a parts list of regulatory elements to an integrated splicing code," *RNA* 14 (2008): 802–813. Freely accessible (2011) at http://rnajournal.cshlp.org/content/14/5/802.full.pdf+html

34. John W. S. Brown, David F. Marshall & Manuel Echeverria, "Intronic noncoding RNAs and splicing," *Trends in Plant Science* 13 (2008): 335–342.

35. Ji Wen, Akira Chiba & Xiaodong Cai, "Computational identification of tissue-specific alternative splicing elements in mouse genes from RNA-Seq," *Nucleic Acids Research* (August 4, 2010). Freely ac-

cessible (2011) at http://nar.oxfordjournals.org/content/early/2010/08/04/nar.gkq679.full.pdf+html

36. Shengdong Ke & Lawrence A. Chasin, "Intronic motif pairs cooperate across exons to promote pre-mRNA splicing," *Genome Biology* 11 (2010): R84. Freely accessible (2011) at http://genomebiology.com/content/pdf/gb-2010-11-8-r84.pdf

37. Yoseph Barash, John A. Calarco, Weijun Gao, Qun Pan, Xinchen Wang, Ofer Shai, Benjamin J. Blencowe & Brendan J. Frey "Deciphering the splicing code," *Nature* 465 (2010): 53–59.

38. Amir Ali Abbasi, Zissis Paparidis, Sajid Malik, Debbie K. Goode, Heather Callaway, Greg Elgar & Karl-Heinz Grzeschik, "Human GLI3 Intragenic Conserved Non-Coding Sequences Are Tissue-Specific Enhancers," *PLoS One* 2:4 (2007): e366. Freely accessible (2011) at http://www.plosone.org/article/info%3Adoi%2F10.1371%2Fjournal.pone.0000366

39. Rodrigo Louro, Tarik El-Jundi, Helder I. Nakaya, Eduardo M. Reis & Sergio Verjovski-Almeida, "Conserved tissue expression signatures of intronic noncoding RNAs transcribed from human and mouse loci," *Genomics* 92 (2008): 18–25.

40. Marc P. Hoeppner, Simon White, Daniel C. Jeffares & Anthony M. Poole, "Evolutionarily Stable Association of Intronic snoRNAs and microRNAs with Their Host Genes," *Genome Biology and Evolution* 2009 (2009): 420–428. Freely accessible (2011) at http://www.ncbi.nlm.nih.gov/pmc/articles/PMC2817437/pdf/evp045.pdf

41. Luis M. Mendes Soares & Juan Valcárcel, "The expanding transcriptome: the genome as the 'Book of Sand,'" *EMBO Journal* 25 (2006): 923–931. Available online with registration (2011) at http://www.nature.com/emboj/journal/v25/n5/full/7601023a.html

42. Antony Rodriguez, Sam Griffiths-Jones, Jennifer L. Ashurst & Allan Bradley, "Identification of Mammalian MicroRNA Host Genes and Transcription Units," *Genome Research* 14 (2004): 1902–1910. Freely accessible (2011) at http://genome.cshlp.org/content/14/10a/1902.full.pdf+html

43. Scott Baskerville & David P. Bartel, "Microarray profiling of microRNAs reveals frequent coexpression with neighboring miRNAs and host genes," *RNA* 11 (2005): 241–247. Freely accessible (2011) at http://rnajournal.cshlp.org/content/11/3/241.full.pdf+html

44. Young-Kook Kim & V. Narry Kim, "Processing of intronic microRNAs," *EMBO Journal* 26 (2007): 775–783. Freely accessible (2011) at http://www.nature.com/emboj/journal/v26/n3/pdf/7601512a.pdf

45. S. Hani Najafi-Shoushtari, Fjoralba Kristo, Yingxia Li, Toshi Shioda, David E. Cohen, Robert E. Gerszten & Anders M. Näär, "MicroRNA-33 and the SREBP Host Genes Cooperate to Control Cholesterol Homeostasis," *Science* 328 (2010): 1566–1569.

46. Alex Mas Monteys, Ryan M. Spengler, Ji Wan, Luis Tecedor, Kimberly A. Lennox, Yi Xing & Beverly L. Davidson, "Structure and activity of putative intronic miRNA promoters," *RNA* 16 (2010): 495–505. Freely accessible (2011) at http://rnajournal.cshlp.org/content/16/3/495.long

47. Michael Bulger & Mark Groudine, "Enhancers: The abundance and function of regulatory sequences beyond promoters," *Developmental Biology* 339 (2010): 250–257.

48. Shawn P. Grogan, Tsaiwei Olee, Koji Hiraoka & Martin K. Lotz, "Repression of Chondrogenesis through Binding of Notch Signaling Proteins HES-1 and HEY-1 to N-box Domains in the *COL2A1* Enhancer Site," *Arthritis & Rheumatism* 58 (2008): 2754–2763. Freely accessible (2011) at http://www3.interscience.wiley.com/cgi-bin/fulltext/121391302/PDFSTART

49. Christopher J. Ott, Neil P. Blackledge, Jenny L. Kerschner, Shih-Hsing Leir, Gregory E. Crawford, Calvin U. Cotton &Ann Harris, "Intronic enhancers coordinate epithelial-specific looping of the active CFTR locus," *Proceedings of the National Academy of Sciences USA* 106 (2009): 19934–19939. Freely accessible (2011) at http://www.pnas.org/content/106/47/19934.full.pdf+html

50. Hani Alotaibi, Elif Yaman, Domenico Salvatore, Valeria Di Dato, Pelin Telkoparan, Roberto Di Lauro & Uygar H. Tazebay, "Intronic elements in the Na$^+$/I$^-$ symporter gene (NIS) interact with retinoic acid receptors and mediate initiation of transcription," *Nucleic Acids Research* 38 (2010): 3172–3185. Freely accessible (2011) at http://nar.oxfordjournals.org/cgi/reprint/38/10/3172

51. Eric I. Campos & Danny Reinberg, "Histones: annotating chromatin," *Annual Review of Genetics* 43 (2009): 559–599.

52. Natalia Soshnikova & Denis Duboule, "Epigenetic Temporal Control of Mouse Hox Genes in Vivo," *Science* 324 (2009): 1320–1323. Available online with registration (2011) at http://www.sciencemag.org/cgi/content/full/324/5932/1320

53. M. R. Hübner & D. L. Spector, "Chromatin dynamics," *Annual Review of Biophysics* 39 (2010): 471–489.

54. S. A. Lavrov & M. V. Kibanov, "Noncoding RNAs and Chromatin Structure," *Biochemistry (Moscow)* 72 (2007): 1422–1438. Freely accessible (2011) at http://protein.bio.msu.ru/biokhimiya/contents/v72/pdf/bcm_1422.pdf

55. Antonio Rodríguez-Campos & Fernando Azorín, "RNA Is an Integral Component of Chromatin that Contributes to Its Structural Organization," *PLoS One* 2:11 (2007): e1182. Freely accessible (2011) at http://www.plosone.org/article/

info%3Adoi%2F10.1371%2Fjournal.
pone.0001182

56. Barbora Malecová & Kevin V Morris,
"Transcriptional gene silencing through
epigenetic changes mediated by non-coding RNAs," *Current Opinion in Molecular
Therapeutics* 12 (2010): 214–222. Freely
accessible (2011) at http://www.ncbi.nlm.
nih.gov/pmc/articles/PMC2861437/
pdf/nihms195819.pdf

57. Daniel P. Caley, Ryan C. Pink, Daniel
Trujillano & David R. F. Carter, "Long
noncoding RNAs, chromatin, and development," *ScientificWorldJournal* 10 (2010):
90–102.

58. Tanmoy Mondal, Markus Rasmussen,
Gaurav Kumar Pandey, Anders Isaksson
& Chandrasekhar Kanduri, "Character-
ization of the RNA content of chromatin,"
Genome Research 20 (2010): 899–907.
Freely accessible (2011) at http://ge-
nome.cshlp.org/content/20/7/899.full.
pdf+html

59. W. F. Chen, K. H. Low, C. Lim & I. Ed-
ery, "Thermosensitive splicing of a clock
gene and seasonal adaptation," *Cold Spring
Harbor Symposia on Quantitative Biology* 72
(2007): 599–606.

60. Dan Xia, Xinxin Huang & Hong Zhang,
"The temporally regulated transcription
factor sel-7 controls developmental timing
in *C. elegans*," *Developmental Biology* 332
(2009): 246–257.

61. David Gubb, "Intron-Delay and the
Precision of Expression of Homeotic Gene
Products in *Drosophila*," *Developmental
Genetics* 7 (1986): 119–131.

62. Carl S. Thummel, "Mechanisms of Transcriptional Timing in *Drosophila*," *Science*
255 (1992): 39–40.

63. Ian A. Swinburne & Pamela A. Silver,
"Intron Delays and Transcriptional Timing During Development," *Developmental
Cell* 14 (2008): 324–330. Freely accessible
(2011) at http://www.ncbi.nlm.nih.gov/
pmc/articles/PMC2825037/pdf/ni-
hms176861.pdf

5. PSEUDOGENES–NOT SO PSEUDO AFTER ALL

1. C. Jacq, J. R. Miller & G. G. Brownlee, "A
pseudogene structure in 5S DNA of *Xeno-
pus laevis*," *Cell* 12 (1977): 109–120.

2. Nick Proudfoot, "Pseudogenes," *Nature*
286 (1980): 840–841.

3. C. Deborah Wilde, "Pseudogenes,"
Critical Reviews in Biochemistry 19 (1986):
323–352.

4. ZhaoLei Zhang & Mark Gerstein, "Large-
scale analysis of pseudogenes in the human
genome," *Current Opinion in Genetics &
Development* 14 (2004): 328–335.

5. Rajkumar Sasidharan & Mark Gerstein,
"Protein fossils live on as RNA," *Nature* 453
(2008): 729–731.

6. Kenneth R. Miller, *Only a Theory: Evolution and the Battle for America's Soul* (New
York: Viking, 2008), pp. 97–98.

7. Douglas J. Futuyma, *Evolution* (Sunderland, MA: Sinauer Associates, 2005), p.
530.

8. Jerry A. Coyne, *Why Evolution Is True*
(New York: Viking, 2009), pp. 66–67.

9. Richard Dawkins, *The Greatest Show on
Earth: The Evidence for Evolution* (New
York: Free Press, 2009), pp. 332–333.

10. John C. Avise, *Inside the Human Genome:
A Case for Non-Intelligent Design* (Oxford:
Oxford University Press, 2010), p. 115.

11. Ilenia D'Errico, Gemma Gadaleta &
Cecilia Saccone, "Pseudogenes in metazoa:
Origin and features," *Briefings in Functional Genomics and Proteomics* 3 (2004):
157–167. Freely accessible (2011) at http://
bfgp.oxfordjournals.org/cgi/reprint/3/2
/157?view=long&pmid=15355597

12. Thierry Tchénio, Evelyne Segal-Bendird-
jian & Thierry Heidmann, "Generation
of processed pseudogenes in murine cells,"
EMBO Journal 12 (1993): 1487–1497.
Freely accessible (2011) at http://
www.ncbi.nlm.nih.gov/pmc/articles/
PMC413361/pdf/emboj00076-0228.
pdf

13. H.-H. M. Dahl, R. M. Brown, W. M. Hutchison, C. Maragos & G. K. Brown, "A testis-specific form of the human pyruvate dehydrogenase E1 alpha subunit is coded for by an intronless gene on chromosome 4," *Genomics* 8 (1990): 225–232.

14. J. Sorge, E. Gross, C. West & E. Beutlert, "High level transcription of the glucocerebrosidase pseudogene in normal subjects and patients with Gaucher disease," *Journal of Clinical Investigation* 86 (1990): 1137–1141. Freely accessible (2011) at http://www.jci.org/articles/view/114818

15. I. Touitou, Q. Q. Cai & H. Rochefort, "17 beta Hydroxysteroid dehydrogenase 1 'pseudogene' is differentially transcribed: still a candidate for the breast-ovarian cancer susceptibility gene (BRCA1)," *Biochemical and Biophysical Research Communications* 201 (1994): 1327–1332.

16. Cornelia Schmutzler & Hans J. Gross, "Genes, variant genes, and pseudogenes of the human tRNA[Val] gene family are differentially expressed in HeLa cells and in human placenta," *Nucleic Acids Research* 18 (1990): 5001–5008. Freely accessible (2011) at http://www.ncbi.nlm.nih.gov/pmc/articles/PMC332105/pdf/nar00201-0021.pdf

17. Yasemin Kaçar, Hildburg Beier & Hans J. Gross, "The presence of tRNA pseudogenes in mammalia and plants and their absence in yeast may account for different specificities of pre-tRNA processing enzymes," *Gene* 156 (1995): 129–132.

18. Erich T. Boger, James R. Sellers & Thomas B. Friedman, "Human myosin XVBP is a transcribed pseudogene," *Journal of Muscle Research and Cell Motility* 22 (2001): 477–483.

19. Richard J. Cristiano, Sara J. Giordano & Alan W. Steggles, "The Isolation and Characterization of the Bovine Cytochrome b$_5$ Gene, and a Transcribed Pseudogene," *Genomics* 17 (1993):348–354.

20. Rainer Fürbass & Jens Vanselow, "An aromatase pseudogene is transcribed in the bovine placenta," *Gene* 154 (1995): 287–291.

21. D. Aubert, C. Nisanz-Sever & M. Herzog, "Mitochondrial rps14 is a transcribed and edited pseudogene in *Arabidopsis thaliana*," *Plant Molecular Biology* 20 (1992): 1169–1174.

22. V. Quiñones, S. Zanlungo, A. Moenne, I. Gómez, L. Holuigue, S. Litvak & X. Jordana, "The rpl5-rps14-cob gene arrangement in *Solanum tuberosum*: rps14 is a transcribed and unedited pseudogene," *Plant Molecular Biology* 31 (1996) 937–943.

23. Deyou Zheng, Zhaolei Zhang, Paul M. Harrison, John Karro, Nick Carriero & Mark Gerstein, "Integrated pseudogene annotation for human chromosome 22: evidence for transcription," *Journal of Molecular Biology* 349 (2005): 27–45.

24. Paul M. Harrison, Deyou Zheng, Zhaolei Zhang, Nicholas Carriero & Mark Gerstein, "Transcribed processed pseudogenes in the human genome: an intermediate form of expressed retrosequence lacking protein-coding ability," *Nucleic Acids Research* 33 (2005): 2374–2383. Freely accessible (2011) at http://nar.oxfordjournals.org/cgi/content/full/33/8/2374

25. Deyou Zheng, Adam Frankish, Robert Baertsch, Philipp Kapranov, Alexandre Reymond, Siew Woh Choo, Yontao Lu, France Denoeud, Stylianos E. Antonarakis, Michael Snyder, Yijun Ruan, Chia-Lin Wei, Thomas R. Gingeras, Roderic Guigó, Jennifer Harrow & Mark B. Gerstein, "Pseudogenes in the ENCODE regions: Consensus annotation, analysis of transcription, and evolution," *Genome Research* 17 (2007): 839–851. Freely accessible (2011) at http://genome.cshlp.org/content/17/6/839.long

26. Michael J. Chorney, Ikuhisa Swada, Gerald A. Gillespie, Rakesh Srivastava, Julian Pan & Sherman M. Weissman, "Transcription Analysis, Physical Mapping, and

Molecular Characterization of a Nonclassical Human Leukocyte Antigen Class I Gene," *Molecular and Cellular Biology* 10 (1990): 243–253. Freely accessible (2011) at http://mcb.asm.org/cgi/reprint/10/1/243?view=long&pmid=2294403

27. Tuan Nguyen, Roger Sunahara, Adriano Marchese, Hubert H. M. Van Tol, Philip Seeman & Brian F. O'Dowd, "Transcription of a human dopamine D5 pseudogene," *Biochemical and Biophysical Research Communications* 181 (1991): 16–21.

28. Jonathan A. Bard, Stanley P. Nawoschik, Brian F. O'Dowd, Susan R. George, Theresa A. Branchek & Richard L. Weinshank, "The human serotonin 5-hydroxytryptamine1D receptor pseudogene is transcribed," *Gene* 153 (1995): 295–296.

29. Christine Pourcel, Jean Jaubert, Michelle Hadchouel, Xue Wu & Johannes Schweizer, "A new family of genes and pseudogenes potentially expressing testis- and brain-specific leucine zipper proteins in man and mouse," *Gene* 249 (2000): 105–113.

30. Mustapha Kandouz, Andrew Bier, George D. Carystinos, Moulay A. Alaoui-Jamali and Gerald Batist, "Connexin43 pseudogene is expressed in tumor cells and inhibits growth," *Oncogene* 23 (2004): 4763–4770.

31. Markus Koller & Emanuel E. Strehler, "Characterization of an intronless human-calmodulin-like pseudogene," *FEBS Letters* 239 (1998): 121–128.

32. Paul Yaswen, Amy Smoll, Junko Hosoda, Gordon Parry & Martha R. Stampfer, "Protein product of a human intronless calmodulin-like gene shows tissue-specific expression and reduced abundance in transformed cells," *Cell Growth & Differentiation* 3 (1992): 335–345. Freely accessible (2010) at http://cgd.aacrjournals.org/cgi/reprint/3/6/335

33. Pete Jeffs & Michael Ashburner, "Processed pseudogenes in *Drosophila*," *Proceedings of the Royal Society (London) B* 244 (1991): 151–159.

34. Manyuan Long & Charles H. Langley, "Natural Selection and the Origin of *jingwei*, a Chimeric Processed Functional Gene in *Drosophila*," *Science* 260 (1993): 91–95.

35. Evgeniy S. Balakirev & Francisco J. Ayala, "Is Esterase-P Encoded by a Cryptic Pseudogene in *Drosophila melanogaster?*" *Genetics* 144 (1996): 1511–1518. Freely accessible (2011) at http://www.genetics.org/cgi/reprint/144/4/1511

36. M. M. Dumancic, J. G. Oakeshott, R. J. Russell & M. J. Healy, "Characterization of the EstP protein in *Drosophila melanogaster* and its conservation in drosophilids," *Biochemical Genetics* 35 (1997): 251–271.

37. Herman A. Dierick, Julian F. B. Mercer & Thomas W. Glover, "A phosphoglycerate mutase brain isoform (*PGAM 1*) pseudogene is localized within the human Menkes disease gene (*ATP7 A*)," *Gene* 198 (1997): 37–41.

38. Esther Betrán, Wen Wang, Li Jin & Manyuan Long, "Evolution of the Phosphoglycerate Mutase Processed Gene in Human and Chimpanzee Revealing the Origin of a New Primate Gene," *Molecular Biology and Evolution* 19 (2002): 654–663. Freely accessible (2011) at http://mbe.oxfordjournals.org/cgi/content/full/19/5/654

39. Agnès Moreau-Aubry, Soizic Le Guiner, Nathalie Labarrière, Marie-Claude Gesnel, Francine Jotereau & Richard Breathnach, "A Processed Pseudogene Codes for a New Antigen Recognized by a CD8[+] T Cell Clone on Melanoma," *Journal of Experimental Medicine* 191 (2000): 1617–1624. Freely accessible (2011) at http://jem.rupress.org/content/191/9/1617.full.pdf+html

40. Bing-Sen Zhou, David R. Beidler & Yung-Chi Cheng, "Identification of Antisense RNA Transcripts from a Human DNA Topoisomerase I Pseudogene," *Cancer Research* 52 (1992): 4280–4285. Freely

accessible (2011) at http://cancerres.aacr-journals.org/cgi/reprint/52/15/4280

41. Dominique Weil, Mary-Anne Power, Graham C. Webb & Chung Leung Li, "Antisense transcription of a murine *FGFR-3* pseudogene during fetal development," *Gene* 187 (1997): 115–122.

42. Andrew Fire, SiQun Xu, Mary K. Montgomery, Steven A. Kostas, Samuel E. Driver & Craig C. Mello, "Potent and specific genetic interference by double-stranded RNA in *Caenorhabditis elegans*," *Nature* 391 (1998): 806–811.

43. Elisabetta Ullu, Appolinaire Djikeng, Huafang Shi & Christian Tschudi, "RNA interference: advances and questions," *Philosophical Transactions of the Royal Society of London B* 357 (2002): 65–70. Freely accessible (2011) at http://rstb.royalsocietypublishing.org/content/357/1417/65.long

44. Gregory J. Hannon, "RNA interference," *Nature* 418 (2002): 244–251.

45. Gunter Meister & Thomas Tuschl, "Mechanisms of gene silencing by double-stranded RNA," *Nature* 431 (2004): 343–349.

46. Sergei A. Korneev, Ji-Ho Park & Michael O'Shea, "Neuronal Expression of Neural Nitric Oxide Synthase (nNOS) Protein Is Suppressed by an Antisense RNA Transcribed from an NOS Pseudogene," *Journal of Neuroscience* 19 (1999): 7711–7720. Freely accessible (2011) at http://www.jneurosci.org/cgi/content/full/19/18/7711

47. Toshiaki Watanabe, Yasushi Totoki, Atsushi Toyoda, Masahiro Kaneda, Satomi Kuramochi-Miyagawa, Yayoi Obata, Hatsune Chiba, Yuji Kohara, Tomohiro Kono, Toru Nakano, M. Azim Surani, Yoshiyuki Sakaki & Hiroyuki Sasaki, "Endogenous siRNAs from naturally formed dsRNAs regulate transcripts in mouse oocytes," *Nature* 453 (2008): 539–543.

48. Oliver H. Tam, Alexei A. Aravin, Paula Stein, Angelique Girard, Elizabeth P. Murchison, Sihem Chelouﬁ, Emily Hodges, Martin Anger, Ravi Sachidanandam, Richard M. Schultz & Gregory J. Hannon, "Pseudogene-derived small interfering RNAs regulate gene expression in mouse oocytes," *Nature* 453 (2008): 534–538.

49. Xingyi Guo, Zhaolei Zhang, Mark B. Gerstein & Deyou Zheng, "Small RNAs Originated from Pseudogenes: cis- or trans-Acting?" *PLos Computational Biology* 5:7 (2009): e1000449. Freely accessible (2011) at http://www.ploscompbiol.org/article/info%3Adoi%2F10.1371%2Fjournal.pcbi.1000449

50. Shinji Hirotsune, Noriyuki Yoshida, Amy Chen, Lisa Garrett, Fumihiro Sugiyama, Satoru Takahashi, Ken-ichi Yagami, Anthony Wynshaw-Boris & Atsushi Yoshiki, "An expressed pseudogene regulates the messenger-RNA stability of its homologous coding gene," *Nature* 423 (2003): 91–96.

51. Yoshihisa Yano, Rintaro Saito, Noriyuki Yoshida, Atsushi Yoshiki, Anthony Wynshaw-Boris, Masaru Tomita & Shinji Hirotsune, "A new role for expressed pseudogenes as ncRNA: regulation of mRNA stability of its homologous coding gene," *Journal of Molecular Medicine* 82 (2004): 414–422.

52. Ondrej Podlaha and Jianzhi Zhang, "Nonneutral Evolution of the Transcribed Pseudogene *Makorin1-p1* in Mice," *Molecular Biology and Evolution* 21 (2004): 2202–2209. Freely accessible (2011) at http://mbe.oxfordjournals.org/cgi/content/full/21/12/2202

53. Jeannie T. Lee, "Molecular biology: Complicity of gene and pseudogene," *Nature* 423 (2003): 26–28.

54. Todd A. Gray, Alison Wilson, Patrick J. Fortin & Robert D. Nicholls, "The putatively functional *Mkrn1-p1* pseudogene is neither expressed nor imprinted, nor does it regulate its source gene in trans," *Proceedings of the National Academy of Sciences*

USA 103 (2006): 12039–12044. Freely accessible (2011) at http://www.pnas.org/content/103/32/12039.full.pdf+html

55. Satoko Kaneko, Ikuko Aki, Kaoru Tsuda, Kazuyuki Mekada, Kazuo Moriwaki, Naoyuki Takahata & Yoko Satta, "Origin and Evolution of Processed Pseudogenes That Stabilize Functional *Makorin1* mRNAs in Mice, Primates and Other Mammals," *Genetics* 172 (2006): 2421–2429. Freely accessible (2011) at http://www.ncbi.nlm.nih.gov/pmc/articles/PMC1456392/pdf/GEN17242421.pdf

56. José Manuel Franco-Zorrilla, Adrián Valli, Marco Todesco, Isabel Mateos, María Isabel Puga, Ignacio Rubio-Somoza, Antonio Leyva, Detlef Weigel, Juan Antonio García & Javier Paz-Ares, "Target mimicry provides a new mechanism for regulation of microRNA activity," *Nature Genetics* 39 (2007): 1033–1037.

57. Armin P. Piehler, Marit Hellum, Jürgen J. Wenzel, Ellen Kaminski, Kari Bente Foss Haug, Peter Kierulf & Wolfgang E. Kaminski, "The human ABC transporter pseudogene family: Evidence for transcription and gene-pseudogene interference," *BMC Genomics* 9 (2008): 165. Freely accessible (2011) at http://www.biomedcentral.com/1471-2164/9/165

58. Laura Poliseno, Leonardo Salmena, Jiangwen Zhang, Brett Carver, William J. Haveman & Pier Paolo Pandolfi, "A coding-independent function of gene and pseudogene mRNAs regulates tumour biology," *Nature* 465 (2010): 1033–1038.

59. Morimitsu Nishikimi & Kunio Yagi, "Molecular basis for the deficiency in humans of gulonolactone oxidase, a key enzyme for ascorbic acid biosynthesis," *American Journal of Clinical Nutrition* 54 (1991): 1203S-1208S. Freely accessible (2011) at http://www.ajcn.org/cgi/reprint/54/6/1203S

60. Morimitsu Nishikimi, Ryuichi Fukuyama, Sinsei Minoshima, Nobuyoshi Shimizu & Kunio Yagi, "Cloning and

Chromosomal Mapping of the Human Nonfunctional Gene for L-Gulono-gamma-lactone Oxidase, the Enzyme for L-Ascorbic Acid Biosynthesis Missing in Man," *Journal of Biological Chemistry* 269 (1994): 13685–13688. Freely accessible (2011) at http://www.jbc.org/content/269/18/13685.long

61. Örjan Svensson, Lars Arvestad & Jens Lagergren, "Genome-Wide Survey for Biologically Functional Pseudogenes," *PLoS Computational Biology* 2:5 (2006): e46. Freely accessible (2011) at http://www.ploscompbiol.org/article/info%3Adoi%2F10.1371%2Fjournal.pcbi.0020046

62. Evgeniy S. Balakirev & Francisco J. Ayala, "Pseudogenes: Are They 'Junk' or Functional DNA?" *Annual Review of Genetics* 37 (2003): 123–51.

63. Amit N. Khachane & Paul M. Harrison, "Assessing the genomic evidence for conserved transcribed pseudogenes under selection," *BMC Genomics* 10 (2009): 435. Freely accessible (2011) at http://www.biomedcentral.com/1471-2164/10/435

6. JUMPING GENES AND REPETITIVE DNA

1. Barbara McClintock, "The Origin and Behavior of Mutable Loci in Maize," *Proceedings of the National Academy of Sciences USA* 36 (1950): 344–355. Freely accessible (2011) at http://www.pnas.org/content/36/6/344.full.pdf+html

2. Georgii P. Georgiev, "Mobile genetic elements in animal cells and their biological significance," *European Journal of Biochemistry* 145 (1984): 203–220. Freely accessible (2011) at http://www3.interscience.wiley.com/cgi-bin/fulltext/120761888/PDFSTART

3. Nina Fedoroff, "How jumping genes were discovered," *Nature Structural Biology* 8 (2001): 300–301. Freely accessible (2011) at http://www.nature.com/nsmb/journal/v8/n4/full/nsb0401_300.html

4. David Baltimore, "Viral RNA-dependent DNA Polymerase: RNA-dependent DNA Polymerase in Virions of RNA Tumour Viruses," *Nature* 226 (1970): 1209–1211.

5. Howard M. Temin & Satoshi Mizutani, "Viral RNA-dependent DNA Polymerase: RNA-dependent DNA Polymerase in Virions of Rous Sarcoma Virus," *Nature* 226 (1970): 1211–1213.

6. P. M. B. Walker & Anne McLaren, "Fractionation of mouse deoxyribonucleic acid on hydroxyapatite," *Nature* 208 (1965): 1175–1179.

7. Roy J. Britten & D. E. Kohne, "Repeated Sequences in DNA," *Science* 161 (1968): 529–540.

8. W. G. Flamm, "Highly Repetitive Sequences of DNA in Chromosomes," *International Review of Cytology* 32 (1972): 1–51.

9. Maxine F. Singer, "SINEs and LINEs: Highly Repeated Short and Long Interspersed Sequences in Mammalian Genomes," *Cell* 28 (1982): 433–434.

10. Bruce Alberts, Alexander Johnson, Julian Lewis, Martin Raff, Keith Roberts & Peter Walter, *Molecular Biology of the Cell*, Fourth Edition (New York: Garland Science, 2002), p. 203

11. GenBank, "Homo sapiens RNA, 7SL, cytoplasmic 1 (RN7SL1), small cytoplasmic RNA," NCBI Reference Sequence: NR_002715.1 (28 February 2010). Freely accessible (2011) at http://www.ncbi.nlm.nih.gov/nuccore/NR_002715.1

12. Karen L. Bennett, Robert E. Hill, Dennis F. Pietras, Mary Woodworth-Gutai, Colleen Kane-Haas, Joanna M. Houston, John K. Heath & Nicholas D. Hastie, "Most Highly Repeated Dispersed DNA Families in the Mouse Genome," *Molecular and Cell Biology* 4 (1984): 1561–1571. Freely accessible (2011) at http://mcb.asm.org/cgi/reprint/4/8/1561?view=long&pmid=6208477

13. Geoffrey J. Faulkner, Yasumasa Kimura, Carsten O. Daub, Shivangi Wani, Charles Plessy, Katharine M. Irvine, Kate Schroder, Nicole Cloonan, Anita L. Steptoe, Timo Lassmann, Kazunori Waki, Nadine Hornig, Takahiro Arakawa, Hazuki Takahashi, Jun Kawai, Alistair R. R. Forrest, Harukazu Suzuki, Yoshihide Hayashizaki, David A. Hume, Valerio Orlando, Sean M. Grimmond & Piero Carninci, "The regulated retrotransposon transcriptome of mammalian cells," *Nature Genetics* 41 (2009): 563–571.

14. Amar Kumar & Jeffrey L. Bennetzen, "Retrotransposons: central players in the structure, evolution and function of plant genomes," *Trends in Plant Science* 5 (2000): 509–510.

15. Hidenori Nishihara, Arian F. A. Smit and Norihiro Okada "Functional noncoding sequences derived from SINEs in the mammalian genome," *Genome Research* 16 (2006): 864–874. Freely accessible (2011) at http://genome.cshlp.org/content/16/7/864.full.pdf+html

16. Craig B. Lowe, Gill Bejerano & David Haussler, "Thousands of human mobile element fragments undergo strong purifying selection near developmental genes," *Proceedings of the National Academy of Sciences USA* 104 (2007): 8005–8010. Freely accessible (2011) at http://www.pnas.org/content/104/19/8005.full.pdf+html

17. Aristotelis Tsirigos & Isidore Rigoutsos, "*Alu* and B1 Repeats Have Been Selectively Retained in the Upstream and Intronic Regions of Genes of Specific Functional Classes," *PLoS Computational Biology* 5:12 (2009): e1000610. Freely accessible (2011) at http://www.ploscompbiol.org/article/info%3Adoi%2F10.1371%2Fjournal.pcbi.1000610

18. Susumu Ohno, W. D. Kaplan & R. Kinosita, "Formation of the sex chromatin by a single X-chromosome in liver cells of *Rattus norvegicus*," *Experimental Cell Research* 18 (1959): 415–418.

19. Jeffrey A. Bailey, Laura Carrel, Aravinda Chakravarti & Evan E. Eichler, "Molecular evidence for a relationship between

LINE-1 elements and X chromosome inactivation: the Lyon repeat hypothesis," *Proceedings of the National Academy of Sciences USA* 97 (2000): 6634–6639. Freely accessible (2011) at http://www.pnas.org/content/97/12/6634.full.pdf+html

20. Y. Amy Tang, Derek Huntley, Giovanni Montana, Andrea Cerase, Tatyana B. Nesterova & Neil Brockdorff, "Efficiency of Xist-mediated silencing on autosomes is linked to chromosomal domain organization," *Epigenetics & Chromatin* 3 (2010): 10. Freely accessible (2011) at http://www.epigeneticsandchromatin.com/content/pdf/1756-8935-3-10.pdf

21. Jennifer C. Chow, Constance Ciaudo, Melissa J. Fazzari. Nathan Mise, Nicolas Servant, Jacob L. Glass, Matthew Attreed, Philip Avner, Anton Wutz, Emmanuel Barillot, John M. Greally, Olivier Voinnet & Edith Heard, "LINE-1 activity in facultative heterochromatin formation during X chromosome inactivation," *Cell* 141 (2010): 956–969.

22. Tammy A. Morrish, Nicolas Gilbert, Jeremy S. Myers, Bethaney J. Vincent, Thomas D. Stamato, Guillermo E. Taccioli, Mark A. Batzer & John V. Moran, "DNA repair mediated by endonuclease-independent LINE-1 retrotransposition," *Nature Genetics* 31 (2002): 159–165.

23. José L. Garcia-Perez, Aurélien J. Doucet, Alain Bucheton, John V. Moran & Nicolas Gilbert, "Distinct mechanisms for trans-mediated mobilization of cellular RNAs by the LINE-1 reverse transcriptase," *Genome Research* 17 (2007): 602–611. Freely accessible (2011) at http://genome.cshlp.org/content/17/5/602.full.pdf+html

24. Elizabeth A. Shepard, Pritpal Chandan, Milena Stevanovic-Walker, Mina Edwards & Ian R. Phillips, "Alternative promoters and repetitive DNA elements define the species-dependent tissue-specific expression of the FMO1 genes of human and mouse," *Biochemical Journal* 406 (2007): 491–499. Freely accessible

(2011) at http://www.biochemj.org/bj/406/0491/4060491.pdf

25. Corrado Spadafora, "A reverse transcriptase-dependent mechanism plays central roles in fundamental biological processes," *Systems Biology in Reproductive Medicine* 54 (2008): 11–21.

26. Anderly C. Chueh, Emma L. Northrop, Kate H. Brettingham-Moore, K. H. Andy Choo & Lee H. Wong, "LINE Retrotransposon RNA Is an Essential Structural and Functional Epigenetic Component of a Core Neocentromeric Chromatin," *PLoS Genetics* 5:1 (2009): e1000354. Freely accessible (2011) at http://www.plosgenetics.org/article/info%3Adoi%2F10.1371%2Fjournal.pgen.1000354

27. Dylan R. Edwards, Craig L. J. Parfett & David T. Denhardt, "Transcriptional Regulation of Two Serum-induced RNAs in Mouse Fibroblasts: Equivalence of One Species to B2 Repetitive Elements," *Molecular and Cellular Biology* 5 (1985): 3280–3288. Freely accessible (2011) at http://mcb.asm.org/cgi/reprint/5/11/3280?view=long&pmid=3837843

28. Roy J. Brittten, "Coding sequences of functioning human genes derived entirely from mobile element sequences," *Proceedings of the National Academy of Sciences USA* 101 (2004): 16825–16830. Freely accessible (2011) at http://www.pnas.org/content/101/48/16825.full.pdf+html

29. L. P. Yavachev, O. I. Georgiev, E. A. Braga, T. A. Avdonina, A. E. Bogomolova, V. B. Zhurkin, V. V. Nosikov & A. A. Hadjiolov, "Nucleotide sequence analysis of the spacer regions flanking the rat rRNA transcription unit and identification of repetitive elements," *Nucleic Acids Research* 14 (1986): 2799–2810. Freely accessible (2011) at http://www.ncbi.nlm.nih.gov/pmc/articles/PMC339699/pdf/nar00275-0387.pdf

30. Richard H. Kimura, Prabhakara V. Choudary & Carl W. Schmid, "Silk worm

Bm1 SINE RNA increases following cellular insults," *Nucleic Acids Research* 27 (1999): 3380–3387. Freely accessible (2011) at http://nar.oxfordjournals.org/cgi/reprint/27/16/3380

31. Richard H. Kimura, Prabhakara V. Choudary, Koni K. Stone & Carl W. Schmid, "Stress induction of Bm1 RNA in silkworm larvae: SINEs, an unusual class of stress genes," *Cell Stress & Chaperones* 6 (2001): 263–272. Freely accessible (2011) at http://www.ncbi.nlm.nih.gov/pmc/articles/PMC434408/pdf/i1466-1268-6-3-263.pdf

32. Phillip A. Yates, Robert W. Burman, Padmaja Mummaneni, Sandra Krussel & Mitchell S. Turker, "Tandem B1 Elements Located in a Mouse Methylation Center Provide a Target for *de Novo* DNA Methylation," *Journal of Biological Chemistry* 274 (1999): 36357–36361. Freely accessible (2011) at http://www.jbc.org/content/274/51/36357.full.pdf+html

33. Celso A. Espinoza, Tiffany A. Allen, Aaron R. Hieb, Jennifer F. Kugel & James A. Goodrich, "B2 RNA binds directly to RNA polymerase II to repress transcript synthesis," *Nature Structural & Molecular Biology* 11 (2004): 822–829.

34. Celso A. Espinoza, James A. Goodrich & Jennifer F. Kugel, "Characterization of the structure, function, and mechanism of B2 RNA, an ncRNA repressor of RNA polymerase II transcription," *RNA* 13 (2007): 583–596. Freely accessible (2011) at http://rnajournal.cshlp.org/content/13/4/583.full.pdf+html

35. Gordon Vansant & Wanda F. Reynolds, "The consensus sequence of a major *Alu* subfamily contains a functional retinoic acid response element," *Proceedings of the National Academy of Sciences USA* 92 (1995): 8229–8233. Freely accessible (2011) at http://www.pnas.org/content/92/18/8229.full.pdf+html

36. Peter D. Mariner, Ryan D. Walters, Celso A. Espinoza, Linda F. Drullinger, Sta-

cey D. Wagner, Jennifer F. Kugel & James A. Goodrich, "Human *Alu* RNA is a modular transacting repressor of mRNA transcription during heat shock," *Molecular Cell* 29 (2008): 499–509.

37. Julien Häsler & Katharina Strub, "*Alu* RNP and *Alu* RNA regulate translation initiation in vitro," *Nucleic Acids Research* 34 (2006): 2374–2385. Freely accessible (2011) at http://nar.oxfordjournals.org/cgi/reprint/34/8/2374

38. Julien Häsler & Katharina Strub, "*Alu* elements as regulators of gene expression," *Nucleic Acids Research* 34 (2006): 5491–5497. Freely accessible (2011) at http://nar.oxfordjournals.org/cgi/content/full/34/19/5491

39. Julien Häsler, T. Samuelsson & Katharina Strub, "Useful 'junk': *Alu* RNAs in the human transcriptome," *Cellular and Molecular Life Sciences* 64 (2007): 1793–1800.

40. Tong J. Gu, Xiang Yi, Xi W. Zhao, Yi Zhao & James Q. Yin, "*Alu*-directed transcriptional regulation of some novel miRNAs," *BMC Genomics* 10 (2009): 563. Freely accessible (2011) at http://www.biomedcentral.com/1471-2164/10/563

41. Michal Barak, Erez Y. Levanon, Eli Eisenberg, Nurit Paz, Gideon Rechavi, George M. Church & Ramit Mehr, "Evidence for large diversity in the human transcriptome created by *Alu* RNA editing," *Nucleic Acids Research* 37 (2009): 6905–6915. Freely accessible (2011) at http://nar.oxfordjournals.org/cgi/reprint/37/20/6905

42. Ryan D. Walters, Jennifer F. Kugel & James A. Goodrich, "InvAluable junk: the cellular impact and function of *Alu* and B2 RNAs," *IUBMB Life* 61 (2009): 831–837.

43. Ann L. Boyle, S. Gwyn Ballard & David C. Ward, "Differential distribution of long and short interspersed element sequences in the mouse genome: Chromosome karyotyping by fluorescence in situ hybridization," *Proceedings of the National Academy of Sciences USA* 87 (1990): 7757–

7761. Freely accessible (2011) at http://www.pnas.org/content/87/19/7757.full.pdf+html

44. J. M. Craig & W. A. Bickmore, "Chromosome bands—flavours to savour," *BioEssays* 15 (1993): 349–354.

45. Yataro Daigo, Minoru Isomura, Tadashi Nishiwaki, Kazufumi Suzuki, Osamu Maruyama, Kumiko Takeuchi, Yuka Yamane, Rie Hayashi, Maiko Minami, Yoshiaki Hojo, Ikuo Uchiyama, Toshihisa Takagi & Yusuke Nakamura, "Significant Differences in the Frequency of Transcriptional Units, Types and Numbers of Repetitive Elements, GC content, and the Number of CpG Islands Between a 1010-kb G-band Genomic Segment on Chromosome 9q31.3 and a 1200-kb R-band Genomic Segment on Chromosome 3p21.3," *DNA Research* 6 (1999): 227–233. Freely accessible (2011) at http://dnaresearch.oxfordjournals.org/cgi/reprint/6/4/227?view=long&pmid=10492169

46. Sam Janssen, Olivier Cuvier, Martin Müller & Ulrich K Laemmli, "Specific gain- and loss-of-function phenotypes induced by satellite-specific DNA-binding drugs fed to *Drosophila melanogaster*," *Molecular Cell* 6 (2000): 1013–1024.

47. Steven Henikoff & Danielle Vermaak, "Bugs on drugs go GAGAA," *Cell* 103 (2000): 695–698.

48. Mary-Lou Pardue & P. Gregory DeBaryshe, "*Drosophila* telomeres: two transposable elements with important roles in chromosomes," *Genetica* 107 (1999): 189–196.

49. Mary-Lou Pardue & P. Gregory DeBaryshe, "*Drosophila* telomere transposons: genetically active elements in heterochromatin," *Genetica* 109 (2000): 45–52.

50. M.-L. Pardue, S. Rashkova, E. Casacuberta, P. G. DeBaryshe, J. A. George & K. L. Traverse, "Two retrotransposons maintain telomeres in *Drosophila*," *Chromosome Research* 13 (2005): 443–453. Freely accessible (2011) at http://www.ncbi.nlm.nih.

gov/pmc/articles/PMC1255937/pdf/nihms3265.pdf

51. Zachary Lippman, Anne-Valérie Gendrel, Michael Black, Matthew W. Vaughn, Neilay Dedhia, W. Richard McCombie, Kimberly Lavine, Vivek Mittal, Bruce May, Kristin D. Kasschau, James C. Carrington, Rebecca W. Doerge, Vincent Colot & Rob Martienssen, "Role of transposable elements in heterochromatin and epigenetic control," *Nature* 430 (2004): 471–476.

52. Zachary Lippman & Rob Martienssen, "The role of RNA interference in heterochromatic silencing," *Nature* 431 (2004): 364–370.

53. Victoria V. Lunyak, Gratien G. Prefontaine, Esperanza Núñez, Thorsten Cramer, Bong-Gun Ju, Kenneth A. Ohgi, Kasey Hutt, Rosa Roy, Angel García-Díaz, Xiaoyan Zhu, Yun Yung, Lluís Montoliu, Christopher K. Glass & Michael G. Rosenfeld, "Developmentally regulated activation of a SINE B2 repeat as a domain boundary in organogenesis," *Science* 317 (2007): 248–251.

54. Ram Parikshan Kumar, Ramamoorthy Senthilkumar, Vipin Singh & Rakesh K. Mishra, "Repeat performance: how do genome packaging and regulation depend on simple sequence repeats?" *BioEssays* 32 (2010): 165–174.

55. Satoru Ide, Takaaki Miyazaki, Hisaji Maki & Takehiko Kobayashi, "Abundance of ribosomal RNA gene copies maintains genome integrity," *Science* 327 (2010): 693–696.

56. Karen Bohmert, Isabelle Camus, Catherine Bellini, David Bouchez, Michel Caboche & Christoph Benning, "*AGO1* defines a novel locus of *Arabidopsis* controlling leaf development," *EMBO Journal* 17 (1998): 170–180. Freely accessible (2011) at http://www.ncbi.nlm.nih.gov/pmc/articles/PMC1170368/pdf/000170.pdf

57. Michelle A. Carmell, Zhenyu Xuan, Michael Q. Zhang & Gregory J. Hannon, "The Argonaute family: tentacles that

reach into RNAi, developmental control, stem cell maintenance, and tumorigenesis," *Genes & Development* 16 (2002): 2733–2742. Freely accessible (2011) at http://genesdev.cshlp.org/content/16/21/2733.full.pdf+html

58. Andreas Lingel & Elisa Izaurralde, "RNAi: finding the elusive endonuclease." *RNA* 10 (2004): 1675–1679. Freely accessible (2011) at http://rnajournal.cshlp.org/content/10/11/1675.full.pdf+html

59. T. Hall, "Structure and function of argonaute proteins," *Structure* 13 (2005): 1403–1408.

60. George L. Sen & Helen M. Blau, "Argonaute 2/RISC resides in sites of mammalian mRNA decay known as cytoplasmic bodies," *Nature Cell Biology* 7 (2005): 633–636.

61. Tim A. Rand, Sean Petersen, Fenghe Du & Xiaodong Wang, "Argonaute2 cleaves the anti-guide strand of siRNA during RISC activation," *Cell* 123 (2005): 621–629.

62. Robert E. Collins & Xiaodong Cheng, "Structural and Biochemical Advances in Mammalian RNAi," *Journal of Cellular Biochemistry* 99 (2006): 1251–1266. Freely accessible (2011) at http://www.ncbi.nlm.nih.gov/pmc/articles/PMC2688788/pdf/nihms-117419.pdf

63. Emily Bernstein, Amy A. Caudy, Scott M. Hammond & Gregory J. Hannon, "Role for a bidentate ribonuclease in the initiation step of RNA interference," *Nature* 409 (2001): 363–366.

64. N. Baumberger & D. C. Baulcombe, "*Arabidopsis* ARGONAUTE1 is an RNA Slicer that selectively recruits microRNAs and short interfering RNAs," *Proceedings of the National Academy of Sciences USA* 102 (2005): 11928–11933. Freely accessible (2011) at http://www.pnas.org/content/102/33/11928.full.pdf+html

65. Haifan Lin & Allan C. Spradling, "A novel group of *pumilio* mutations affects the asymmetric division of germline stem cells in the *Drosophila* ovary," *Development* 124 (1997): 2463–2476. Freely accessible (2011) at http://dev.biologists.org/content/124/12/2463.long

66. Daniel N. Cox, Anna Chao & Haifan Lin, "*piwi* encodes a nucleoplasmic factor whose activity modulates the number and division rate of germline stem cells," *Development* 127 (2000): 503–514. Freely accessible (2011) at http://dev.biologists.org/content/127/3/503.long

67. Daniel N. Cox, Anna Chao, Jeff Baker, Lisa Chang, Dan Qiao & Haifan Lin, "A novel class of evolutionarily conserved genes defined by *piwi* are essential for stem cell self-renewal," *Genes & Development* 12 (1998): 3715–3727. Freely accessible (2011) at http://genesdev.cshlp.org/content/12/23/3715.full.pdf+html

68. Jin-Biao Ma, Yu-Ren Yuan, Gunter Meister, Yi Pei, Thomas Tuschl & Dinshaw J. Patel, "Structural basis for 5'-end-specific recognition of guide RNA by the *A. fulgidus* Piwi protein," *Nature* 434 (2005): 666–670.

69. Christopher R. Faehnle & Leemor Joshua-Tor, "Argonautes confront new small RNAs," *Current Opinion in Chemical Biology* 11 (2007): 569–577. Freely accessible (2011) at http://www.ncbi.nlm.nih.gov/pmc/articles/PMC2077831/pdf/nihms-33482.pdf

70. Julia Höck & Gunter Meister, "The Argonaute protein family," *Genome Biology* 9:2 (2008): 210. Freely accessible (2011) at http://genomebiology.com/content/pdf/gb-2008-9-2-210.pdf

71. Kuniaki Saito, Kazumichi M. Nishida, Tomoko Mori, Yoshinori Kawamura, Keita Miyoshi, Tomoko Nagami, Haruhiko Siomi & Mikiko C. Siomi, "Specific association of Piwi with rasiRNAs derived from retrotransposon and heterochromatic regions in the *Drosophila* genome," *Genes & Development* 20 (2006): 2214–2222. Freely accessible (2011) at http://genes-

dev.cshlp.org/content/20/16/2214.full.
pdf+html

72. Yoshinori Kawamura, Kuniaki Saito, Taishin Kin, Yukiteru Ono, Kiyoshi Asai, Takafumi Sunohara, Tomoko N. Okada, Mikiko C. Siomi & Haruhiko Siomi, "Drosophila endogenous small RNAs bind to Argonaute 2 in somatic cells," Nature 453 (2008): 793–797.

73. Megha Ghildiyal, Hervé Seitz, Michael D. Horwich, Chengjian Li, Tingting Du, Soohyun Lee, Jia Xu, Ellen L.W. Kittler, Maria L. Zapp, Zhiping Weng & Phillip D. Zamore, "Endogenous siRNAs Derived from Transposons and mRNAs in Drosophila Somatic Cells," Science 320 (2008): 1077–1081.

74. Haifan Lin, "piRNAs in the germ line," Science 316 (2007): 397.

75. Travis Thomson & Haifan Lin, "The Biogenesis and Function of PIWI Proteins and piRNAs: Progress and Prospect," Annual Review of Cell and Developmental Biology 25 (2009): 355–376. Freely accessible (2011) at http://www.ncbi.nlm.nih.gov/pmc/articles/PMC2780330/pdf/nihms158620.pdf

76. Christel Rouget, Catherine Papin, Anthony Boureux, Anne-Cécile Meunier, Bénédicte Franco, Nicolas Robine, Eric C. Lai, Alain Pelisson & Martine Simonelig, "Maternal mRNA deadenylation and decay by the piRNA pathway in the early Drosophila embryo," Nature 467 (2010): 1128–1132.

77. Howard M. Temin, "Homology Between RNA from Rous Sarcoma Virus and DNA from Rous Sarcoma Virus-Infected Cells," Proceedings of the National Academy of Sciences USA 52 (1964): 323–329. Freely accessible (2011) at http://www.pnas.org/content/52/2/323.full.pdf+html

78. Heiner Westphal & Renato Dulbecco, "Viral DNA in Polyoma- and SV40-transformed Cell Lines," Proceedings of the National Academy of Sciences USA 59 (1968): 1158–1165. Freely accessible

(2011) at http://www.pnas.org/content/59/4/1158.full.pdf+html

79. Harold E. Varmus "Retroviruses," Science 240 (1988): 1427–1435.

80. Peter N. Rosenthal, Harriet L. Robinson, William S. Robinson, Teruko Hanafusa & Hidesaburo Hanafusa, "DNA in Uninfected and Virus-Infected Cells Complementary to Avian Tumor Virus RNA," Proceedings of the National Academy of Sciences USA 68 (1971): 2336–2340. Freely accessible (2011) at http://www.pnas.org/content/68/10/2336.full.pdf+html

81. Harold E. Varmus, Robin A. Weiss, Robert R. Friis, Warren Levinson & J. Michael Bishop, "Detection of Avian Tumor Virus-Specific Nucleotide Sequences in Avian Cell DNAs," Proceedings of the National Academy of Sciences USA 69 (1972): 20–24. Freely accessible (2011) at http://www.pnas.org/content/69/1/20.full.pdf+html

82. Robin A Weiss, "The discovery of endogenous retroviruses," Retrovirology 3 (2006): 67. Freely accessible (2011) at http://www.retrovirology.com/content/pdf/1742-4690-3-67.pdf

83. Motoharu Seiki, Seisuke Hattori & Mistuaki Yoshida, "Human adult T-cell leukemia virus: Molecular cloning of the provirus DNA and the unique terminal structure," Proceedings of the National Academy of Sciences USA 79 (1982): 6899–6902. Freely accessible (2011) at http://www.pnas.org/content/79/22/6899.full.pdf+html

84. Joseph G. Sodroski, Craig A. Rosen & William A. Haseltine, "Trans-acting transcriptional activation of the long terminal repeat of human T lymphotropic viruses in infected cells," Science 225 (1984): 381–385.

85. Robert G. Ramsay, Shunsuke Ishii & Thomas J. Gonda, "Interaction of the Myb Protein with Specific DNA Binding Sites," Journal of Biological Chemistry 267 (1992): 5656–5662. Freely accessible

(2011) at http://www.jbc.org/content/267/8/5656.full.pdf+html

86. Nathalie de Parseval, Hanan Alkabbani & Thierry Heidmann, "The long terminal repeats of the HERV-H human endogenous retrovirus contain binding sites for transcriptional regulation by the Myb protein," *Journal of General Virology* 80 (1999): 841–845. Freely accessible (2011) at http://vir.sgmjournals.org/cgi/reprint/80/4/841

87. Patrik Medstrand, Josette-Renée Landry & Dixie L. Mager, "Long Terminal Repeats Are Used as Alternative Promoters for the Endothelin B Receptor and Apolipoprotein C-I Genes in Humans," *Journal of Biological Chemistry* 276 (2001): 1896–1903. Freely accessible (2011) at http://www.jbc.org/content/276/3/1896.full.pdf+html

88. Josette-Renée Landry & Dixie L. Mager, "Functional Analysis of the Endogenous Retroviral Promoter of the Human Endothelin B Receptor Gene," *Journal of Virology* 77 (2003): 7459–7466. Freely accessible (2011) at http://jvi.asm.org/cgi/reprint/77/13/7459

89. Anne E. Peaston, Alexei V. Evsikov, Joel H. Graber, Wilhelmine N. de Vries, Andrea E. Holbrook, Davor Solter & Barbara B. Knowles, "Retrotransposons regulate host genes in mouse oocytes and preimplantation embryos," *Developmental Cell* 7 (2004): 597–606.

90. James A. Shapiro, "Retrotransposons and regulatory suites," *BioEssays* 27 (2005): 122–125.

91. Jianhua Ling, Wenhu Pi, Roni Bollag, Shan Zeng, Meral Keskintepe, Hatem Saliman, Sanford Krantz, Barry Whitney & Dorothy Tuan, "The Solitary Long Terminal Repeats of ERV-9 Endogenous Retrovirus Are Conserved during Primate Evolution and Possess Enhancer Activities in Embryonic and Hematopoietic Cells," *Journal of Virology* 76 (2002): 2410–2423.

Freely accessible (2011) at http://jvi.asm.org/cgi/reprint/76/5/2410

92. Elena Gogvadze, Elena Stukacheva, Anton Buzdin & Eugene Sverdlov, "Human-Specific Modulation of Transcriptional Activity Provided by Endogenous Retroviral Insertions," *Journal of Virology* 83 (2009): 6098–6105. Freely accessible (2011) at http://jvi.asm.org/cgi/reprint/83/12/6098

93. Catherine A. Dunn, Patrik Medstrand & Dixie L. Mager, "An endogenous retroviral long terminal repeat is the dominant promoter for human β1,3-galactosyltransferase 5 in the colon," *Proceedings of the National Academy of Sciences USA* 100 (2003): 12841–12846. Freely accessible (2011) at http://www.pnas.org/content/100/22/12841.full.pdf+html

94. Catherine A. Dunn & Dixie L. Mager, "Transcription of the human and rodent SPAM1 / PH-20 genes initiates within an ancient endogenous retrovirus," *BMC Genomics* 6 (2005): 47. Freely accessible (2011) at http://www.biomedcentral.com/content/pdf/1471-2164-6-47.pdf

95. Anton Buzdin, Elena Kovalskaya-Alexandrova, Elena Gogvadze & Eugene Sverdlov, "At Least 50% of Human-Specific HERV-K (HML-2) Long Terminal Repeats Serve in Vivo as Active Promoters for Host Nonrepetitive DNA Transcription," *Journal of Virology* 80 (2006): 10752–10762. Freely accessible (2011) at http://jvi.asm.org/cgi/reprint/80/21/10752

96. Woo Jung Lee, Hyun Jin Kwun & Kyung Lib Jang, "Analysis of transcriptional regulatory sequences in the human endogenous retrovirus W long terminal repeat," *Journal of General Virology* 84 (2003): 2229–2235. Freely accessible (2011) at http://vir.sgmjournals.org/cgi/reprint/84/8/2229

97. Andrew B. Conley, Jittima Piriyapongsa & I. King Jordan, "Retroviral promoters in the human genome," *Bioinformatics* 24 (2008): 1563–1567. Freely accessible

(2011) at http://bioinformatics.oxford-journals.org/cgi/reprint/24/14/1563

98. Ulrike Schön, Olivia Diem, Laura Leitner, Walter H. Günzburg, Dixie L. Mager, Brian Salmons & Christine Leib-Mösch, "Human Endogenous Retroviral Long Terminal Repeat Sequences as Cell Type-Specific Promoters in Retroviral Vectors," *Journal of Virology* 83 (2009): 12643–12650. Freely accessible (2011) at http://jvi.asm.org/cgi/reprint/83/23/12643

99. Patrick J. W. Venables, Sharon M. Brookes, David Griffiths, Robin A. Weiss & Mark T. Boyd, "Abundance of an endogenous retroviral envelope protein in placental trophoblasts suggests a biological function." *Virology* 211 (1995): 589–592.

100. Sha Mi, Xinhua Lee, Xiang-ping Li, Geertruida M. Veldman, Heather Finnerty, Lisa Racie, Edward LaVallie, Xiang-Yang Tang, Philippe Edouard, Steve Howes, James C. Keith Jr. & John M. McCoy, "Syncytin is a captive retroviral envelope protein involved in human placental morphogenesis," *Nature* 403 (2000): 785–789.

101. Jean-Luc Blond, Dimitri Lavillette, Valérie Cheynet, Olivier Bouton, Guy Oriol, Sylvie Chapel-Fernandes, Bernard Mandrand, François Mallet & François-Loïc Cosset, "An Envelope Glycoprotein of the Human Endogenous Retrovirus HERV-W Is Expressed in the Human Placenta and Fuses Cells Expressing the Type D Mammalian Retrovirus Receptor," *Journal of Virology* 74 (2000): 3321–3329. Freely accessible (2011) at http://jvi.asm.org/cgi/reprint/74/7/3321

102. Jean-Louis Frendo, Delphine Olivier, Valérie Cheynet, Jean-Luc Blond, Olivier Bouton, Michel Vidaud, Michèle Rabreau, Danièle Evain-Brion & François Mallet, "Direct Involvement of HERV-W Env Glycoprotein in Human Trophoblast Cell Fusion and Differentiation," *Molecular and Cellular Biology* 23 (2003): 3566–3574.

Freely accessible (2011) at http://mcb.asm.org/cgi/reprint/23/10/3566

103. I. Knerr, B. Huppertz, C. Weigel, J. Dötsch, C. Wich, R. L. Schild, M. W. Beckmann & W. Rascher, "Endogenous retroviral syncytin: compilation of experimental research on syncytin and its possible role in normal and disturbed human placentogenesis," *Molecular Human Reproduction* 10 (2004): 581–588. Freely accessible (2011) at http://molehr.oxfordjournals.org/cgi/reprint/10/8/581

104. Kathrin A. Dunlap, Massimo Palmarini, Mariana Varela, Robert C. Burghardt, Kanako Hayashi, Jennifer L. Farmer & Thomas E. Spencer, "Endogenous retroviruses regulate periimplantation placental growth and differentiation," *Proceedings of the National Academy of Sciences USA* 103 (2006): 14390–14395. Freely accessible (2011) at http://www.pnas.org/content/103/39/14390.full.pdf+html

105. Ina Knerr, Ernst Beinder & Wolfgang Rascher, "Syncytin, a novel human endogenous retroviral gene in human placenta: Evidence for its dysregulation in preeclampsia and HELLP syndrome," *American Journal of Obstetrics and Gynecology* 186 (2002): 210–213.

106. Joseph M. Antony, Kristofor K. Ellestad, Robert Hammond, Kazunori Imaizumi, Francois Mallet, Kenneth G. Warren & Christopher Power, "The Human Endogenous Retrovirus Envelope Glycoprotein, Syncytin-1, Regulates Neuroinflammation and its Receptor Expression in Multiple Sclerosis: A Role for Endoplasmic Reticulum Chaperones in Astrocytes," *Journal of Immunology* 179 (2007): 1210–1224. Freely accessible (2011) at http://www.jimmunol.org/cgi/reprint/179/2/1210

107. Sandra Blaise, Nathalie de Parseval, Laurence Bénit & Thierry Heidmann, "Genomewide screening for fusogenic human endogenous retrovirus envelopes identifies *syncytin 2*, a gene conserved on primate evolution," *Proceedings of the*

National Academy of Sciences USA 100 (2003): 13013–13018. Freely accessible (2011) at http://www.pnas.org/content/100/22/13013.full.pdf+html

108. Cécile Esnault, Stéphane Priet, David Ribet, Cécile Vernochet, Thomas Bruls, Christian Lavialle, Jean Weissenbach & Thierry Heidmann, "A placenta-specific receptor for the fusogenic, endogenous retrovirus-derived, human syncytin-2," *Proceedings of the National Academy of Sciences USA* 105 (2008): 17532–17537. Freely accessible (2011) at http://www.pnas.org/content/105/45/17532.full.pdf+html

109. Anne Dupressoir, Geoffroy Marceau, Cécile Vernochet, Laurence Bénit, Colette Kanellopoulos, Vincent Sapin & Thierry Heidmann, "Syncytin-A and syncytin-B, two fusogenic placenta-specific murine envelope genes of retroviral origin conserved in Muridae," *Proceedings of the National Academy of Sciences USA* 102 (2005): 725–730. Freely accessible (2011) at http://www.pnas.org/content/102/3/725.full.pdf+html

110. Anne Dupressoir, Cécile Vernochet, Olivia Bawa, Francis Harper, Gérard Pierron, Paule Opolon & Thierry Heidmann, "Syncytin-A knockout mice demonstrate the critical role in placentation of a fusogenic, endogenous retrovirus-derived, envelope gene," *Proceedings of the National Academy of Sciences USA* 106 (2009): 12127–12132. Freely accessible (2011) at http://www.pnas.org/content/106/29/12127.full.pdf+html

111. Odile Heidmann, Cécile Vernochet, Anne Dupressoir & Thierry Heidmann, "Identification of an endogenous retroviral envelope gene with fusogenic activity and placenta-specific expression in the rabbit: a new 'syncytin' in a third order of mammals," *Retrovirology* 6 (2009): 107. Freely accessible (2011) at http://www.retrovirology.com/content/6/1/107

112. Jonathan P. Stoye, "Proviral protein provides placental function," *Proceedings of the National Academy of Sciences USA* 106 (2009): 11827–11828. Freely accessible (2011) at http://www.pnas.org/content/106/29/11827.full.pdf+html

113. Sarah Prudhomme, Guy Oriol & François Mallet, "A Retroviral Promoter and a Cellular Enhancer Define a Bipartite Element which Controls *env* ERVWE1 Placental Expression," *Journal of Virology* 78 (2004): 12157–12168. Freely accessible (2011) at http://jvi.asm.org/cgi/reprint/78/22/12157

114. You-Hong Cheng, Brian D. Richardson, Michael A. Hubert & Stuart Handwerger, "Isolation and Characterization of the Human Syncytin Gene Promoter," *Biology of Reproduction* 70 (2004): 694–701. Freely accessible (2011) at http://www.biolreprod.org/content/70/3/694.full.pdf+html

115. François Mallet, Olivier Bouton, Sarah Prudhomme, Valérie Cheynet, Guy Oriol, Bertrand Bonnaud, Gérard Lucotte, Laurent Duret & Bernard Mandrand, "The endogenous retroviral locus *ERVWE-1* is a bona fide gene involved in hominoid placental physiology," *Proceedings of the National Academy of Sciences USA* 101 (2004): 1731–1736. Freely accessible (2011) at http://www.pnas.org/content/101/6/1731.full.pdf+html

116. James A. Shapiro & Richard von Sternberg, "Why repetitive DNA is essential to genome function," *Biological Reviews* 80 (2005): 227–250. Freely accessible (2011) at http://shapiro.bsd.uchicago.edu/index3.html?content=genome.html

117. Francis S. Collins, *The Language of God: A Scientist Presents Evidence for Belief* (New York: Free Press, 2006), pp. 136–137.

7. FUNCTIONS INDEPENDENT OF EXACT SEQUENCE

1. David Gubb, "Intron-Delay and the Precision of Expression of Homeotic Gene

Products in *Drosophila*," *Developmental Genetics* 7 (1986): 119–131.

2. Carl S. Thummel, "Mechanisms of Transcriptional Timing in *Drosophila*," *Science* 255 (1992): 39–40.

3. Ian A. Swinburne & Pamela A. Silver, "Intron Delays and Transcriptional Timing during Development," *Developmental Cell* 14 (2008): 324–330. Freely accessible (2011) at http://www.ncbi.nlm.nih.gov/pmc/articles/PMC2825037/pdf/nihms176861.pdf

4. Jennifer C. Chow, Constance Ciaudo, Melissa J. Fazzari. Nathan Mise, Nicolas Servant, Jacob L. Glass, Matthew Attreed, Philip Avner, Anton Wutz, Emmanuel Barillot, John M. Greally, Olivier Voinnet & Edith Heard, "LINE-1 activity in facultative heterochromatin formation during X chromosome inactivation," *Cell* 141 (2010): 956–969.

5. Ann L. Boyle, S. Gwyn Ballard & David C. Ward, "Differential distribution of long and short interspersed element sequences in the mouse genome: chromosome karyotyping by fluorescence in situ hybridization," *Proceedings of the National Academy of Sciences USA* 87 (1990): 7757–7761. Freely accessible (2011) at http://www.pnas.org/content/87/19/7757.full.pdf+html

6. J. M. Craig & W. A. Bickmore, "Chromosome bands—flavours to savour," *BioEssays* 15 (1993): 349–354.

7. Yataro Daigo, Minoru Isomura, Tadashi Nishiwaki, Kazufumi Suzuki, Osamu Maruyama, Kumiko Takeuchi, Yuka Yamane, Rie Hayashi, Maiko Minami, Yoshiaki Hojo, Ikuo Uchiyama, Toshihisa Takagi & Yusuke Nakamura, "Significant Differences in the Frequency of Transcriptional Units, Types and Numbers of Repetitive Elements, GC Content, and the Number of CpG Islands Between a 1010-kb G-band Genomic Segment on Chromosome 9q31.3 and a 1200-kb R-band Genomic Segment on Chromosome 3p21.3," *DNA Research* 6 (1999): 227–233. Freely accessible (2011) at http://dnaresearch.oxfordjournals.org/cgi/reprint/6/4/227?view=long&pmid=10492169

8. Roel van Driel, Paul F. Fransz & Pernette J. Verschure, "The eukaryotic genome: a system regulated at different hierarchical levels," *Journal of Cell Science* 116 (2003): 4067–4075. Freely accessible (2011) at http://jcs.biologists.org/cgi/reprint/116/20/4067

9. Tom Misteli, "Beyond the sequence: cellular organization of genome function," *Cell* 128 (2007): 787–800.

10. Emile Zuckerkandl, "Junk DNA and sectorial gene repression," *Gene* 205 (1997): 323–343.

11. Emile Zuckerkandl, "Why so many noncoding nucleotides? The eukaryote genome as an epigenetic machine," *Genetica* 115 (2002): 105–129.

12. Emile Zuckerkandl & Giacomo Cavalli, "Combinatorial epigenetics, 'junk DNA', and the evolution of complex organisms," *Gene* 390 (2007): 232–242.

13. Michael Bulger & Mark Groudine, "Looping versus linking: toward a model for long-distance gene activation," *Genes & Development* 13 (1999): 2465–2477. Freely accessible (2011) at http://genesdev.cshlp.org/content/13/19/2465.full.pdf+html

14. Gong Hong Wei, De Pei Liu & Chih Chuan Liang, "Chromatin domain boundaries: insulators and beyond," *Cell Research* 15 (2005): 292–300. Freely accessible (2011) at http://www.nature.com/cr/journal/v15/n4/pdf/7290298a.pdf

15. Susan E. Celniker & Robert A. Drewell, "Chromatin looping mediates boundary element promoter interactions," *BioEssays* 29 (2007): 7–10.

16. Peter Fraser, "Transcriptional control thrown for a loop," *Current Opinion in Genetics & Development* 16 (2006): 490–495.

17. Robert-Jan T. S. Palstra, "Close encounters of the 3C kind: long-range chromatin

interactions and transcriptional regulation," *Briefings in Functional Genomics and Proteomics* 8 (2009): 297–309.

18. Anita Göndör & Rolf Ohlsson, "Chromosome crosstalk in three dimensions," *Nature* 461 (2009): 212–217.

19. John L. Rinn, Michael Kertesz, Jordon K. Wang, Sharon L. Squazzo, Xiao Xu, Samantha A. Brugmann, Henry Goodnough, Jill A. Helms, Peggy J. Farnham, Eran Segal & Howard Y. Chang, "Functional Demarcation of Active and Silent Chromatin Domains in Human HOX Loci by Noncoding RNAs," *Cell* 129 (2007): 1311–1323. Freely accessible (2011) at http://www.ncbi.nlm.nih.gov/pmc/articles/PMC2084369/pdf/nihms26949.pdf

20. Miao-Chih Tsai, Ohad Manor, Yue Wan, Nima Mosammaparast, Jordon K. Wang, Fei Lan, Yang Shi, Eran Segal & Howard Y. Chang, "Long Noncoding RNA as Modular Scaffold of Histone Modification Complexes," *Science* 329 (2010): 689–693. Freely accessible (2011) at http://www.ncbi.nlm.nih.gov/pmc/articles/PMC2967777/pdf/nihms244741.pdf

21. Michaela Angermayr, Ulrich Oechsner, Kerstin Gregor, Gary P. Schroth & Wolfhard Bandlow, "Transcription initiation in vivo without classical transactivators: DNA kinks flanking the core promoter of the housekeeping yeast adenylate kinase gene, AKY2, position nucleosomes and constitutively activate transcription," *Nucleic Acids Research* 30 (2002): 4199–4207. Freely accessible (2011) at http://nar.oxfordjournals.org/content/30/19/4199.full.pdf+html

22. Jason A. Greenbaum, Bo Pang & Thomas D. Tullius "Construction of a genome-scale structural map at single-nucleotide resolution," *Genome Research* 17 (2007): 947–953. Freely accessible (2011) at http://genome.cshlp.org/content/17/6/947.full.pdf+html

23. Stephen C. J. Parker, Loren Hansen, Hatice Ozel Abaan, Thomas D. Tullius & Elliott H. Margulies, "Local DNA Topography Correlates with Functional Noncoding Regions of the Human Genome," *Science* 324 (2009): 389–392.

24. Antonio Rodríguez-Campos & Fernando Azorín, "RNA Is an Integral Component of Chromatin that Contributes to Its Structural Organization," *PLoS One* 2:11 (2007): e1182. Freely accessible (2011) at http://www.plosone.org/article/info%3Adoi%2F10.1371%2Fjournal.pone.0001182

25. Helder Maiato, Jennifer DeLuca, E. D. Salmon & William C. Earnshaw, "The dynamic kinetochore-microtubule interface," *Journal of Cell Science* 117 (2004): 5461–5477. Freely accessible (2011) at http://jcs.biologists.org/cgi/reprint/117/23/5461

26. Xingkun Liu, Ian McLeod, Scott Anderson, John R. Yates III & Xiangwei He, "Molecular analysis of kinetochore architecture in fission yeast," *EMBO Journal* 24 (2005): 2919–2930. Freely accessible (2011) at http://www.nature.com/emboj/journal/v24/n16/pdf/7600762a.pdf

27. Ajit P. Joglekar, David C. Bouck, Jeffrey N. Molk, Kerry S. Bloom & Edward D. Salmon, "Molecular architecture of a kinetochore-microtubule attachment site," *Nature Cell Biology* 8 (2006): 581–585. Freely accessible (2011) at http://www.ncbi.nlm.nih.gov/pmc/articles/PMC2867088/pdf/nihms199890.pdf

28. Iain M. Cheeseman & Arshad Desai, "Molecular architecture of the kinetochore–microtubule interface," *Nature Reviews Molecular Cell Biology* 9 (2008): 33–46.

29. Xiaohu Wan, Ryan P. O'Quinn, Heather L. Pierce, Ajit P. Joglekar, Walt E. Gall, Jennifer G. DeLuca, Christopher W. Carroll, Song-Tao Liu, Tim J. Yen, Bruce F. McEwen, P. Todd Stukenberg, Arshad Desai & Edward D. Salmon, "Protein Architecture of the Human Kinetochore Microtubule Attachment Site," *Cell*

137 (2009): 672–684. Freely accessible (2011) at http://www.ncbi.nlm.nih.gov/pmc/articles/PMC2699050/pdf/nihms114285.pdf

30. Ajit P. Joglekar, Kerry S. Bloom & Edward D. Salmon, "In vivo protein architecture of the eukaryotic kinetochore with nanometer scale accuracy," *Current Biology* 19 (2009): 694–699. Freely accessible (2011) at http://www.ncbi.nlm.nih.gov/pmc/articles/PMC2832475/pdf/nihms178290.pdf

31. Katheleen Gardiner, "Clonability and gene distribution on human chromosome 21: reflections of junk DNA content?" *Gene* 205 (1997): 39–46.

32. James W. Gaubatz & Richard G. Cutler, "Mouse Satellite DNA Is Transcribed in Senescent Cardiac Muscle," *Journal of Biological Chemistry* 265 (1990): 17753–17758. Freely accessible (2011) at http://www.jbc.org/content/265/29/17753.long

33. F. Rudert, S. Bronner, J. M. Garnier & P. Dollé, "Transcripts from opposite strands of gamma satellite DNA are differentially expressed during mouse development," *Mammalian Genome* 6 (1995): 76–83.

34. Brenda J. Reinhart & David P. Bartel, "Small RNAs Correspond to Centromere Heterochromatic Repeats," *Science* 297 (2002): 1831. Available online with registration (2011) at http://www.sciencemag.org/cgi/content/full/297/5588/1831

35. Bruce P. May, Zachary B. Lippman, Yuda Fang, David L. Spector & Robert A. Martienssen, "Differential Regulation of Strand-Specific Transcripts from *Arabidopsis* Centromeric Satellite Repeats," *PLoS Genetics* 1:6 (2005): e79. Freely accessible (2011) at http://www.ncbi.nlm.nih.gov/pmc/articles/PMC1317654/pdf/pgen.0010079.pdf

36. Junjie Lu & David M. Gilbert, "Proliferation-dependent and cell cycle regulated transcription of mouse pericentric heterochromatin," *Journal of Cell Biology* 179 (2007): 411–412. Freely accessible (2011) at http://jcb.rupress.org/content/179/3/411.full.pdf+html

37. Rachel J. O'Neill & Dawn M. Carone, "The role of ncRNA in centromeres: a lesson from marsupials," *Progress in Molecular and Subcellular Biology* 48 (2009): 77–101.

38. Thomas A. Volpe, Catherine Kidner, Ira M. Hall, Grace Teng, Shiv I. S. Grewal & Robert A. Martienssen, "Regulation of Heterochromatic Silencing and Histone H3 Lysine-9 Methylation by RNAi," *Science* 297 (2002): 1833–1837. Available online with registration (2011) at http://www.sciencemag.org/cgi/content/full/297/5588/1833

39. Christopher N. Topp, Cathy X. Zhong & R. Kelly Dawe, "Centromere-encoded RNAs are integral components of the maize kinetochore," *Proceedings of the National Academy of Sciences USA* 101 (2004): 15986–15991. Freely accessible (2011) at http://www.pnas.org/content/101/45/15986.full.pdf+html

40. Haniaa Bouzinba-Segard, Adeline Guais & Claire Francastel, "Accumulation of small murine minor satellite transcripts leads to impaired centromeric architecture and function," *Proceedings of the National Academy of Sciences USA* 103 (2006): 8709–8714. Freely accessible (2011) at http://www.pnas.org/content/103/23/8709.full.pdf+html

41. Federica Ferri, Haniaa Bouzinba-Segard, Guillaume Velasco, Florent Hubé & Claire Francastel, "Non-coding murine centromeric transcripts associate with and potentiate Aurora B kinase," *Nucleic Acids Research* 37 (2009): 5071–5080. Freely accessible (2011) at http://nar.oxfordjournals.org/cgi/reprint/37/15/5071

42. Lee H. Wong, Kate H. Brettingham-Moore, Lyn Chan, Julie M. Quach, Melissa A. Anderson, Emma L. Northrop, Ross Hannan, Richard Saffery, Margaret L. Shaw, Evan Williams & K. H. Andy Choo, "Centromere RNA is a key component for the assembly of nucleoproteins at

the nucleolus and centromere," *Genome Research* 17 (2007): 1146–1160. Freely accessible (2011) at http://genome.cshlp.org/content/17/8/1146.full.pdf+html

43. Shiv I. S. Grewal & Sarah C. R. Elgin, "Transcription and RNA interference in the formation of heterochromatin," *Nature* 447 (2007): 399–406.

44. Tom Volpe, Vera Schramke, Georgina L. Hamilton, Sharon A. White, Grace Teng, Robert A. Martienssen & Robin C. Allshire, "RNA interference is required for normal centromere function in fission yeast," *Chromosome Research* 11 (2003): 137–146.

45. André Verdel, Songtao Jia, Scott Gerber, Tomoyasu Sugiyama, Steven Gygi, Shiv I. S. Grewal & Danesh Moazed, "RNAi-Mediated Targeting of Heterochromatin by the RITS Complex," *Science* 303 (2004): 672–676. Available online with registration (2011) at http://www.sciencemag.org/cgi/content/full/303/5658/672

46. Mohammad R. Motamedi, André Verdel, Serafin U. Colmenares, Scott A. Gerber, Steven P. Gygi & Danesh Moazed, "Two RNAi complexes, RITS and RDRC, physically interact and localize to noncoding centromeric RNAs," *Cell* 119 (2004): 789–802.

47. Hiroaki Kato, Derek B. Goto, Robert A. Martienssen, Takeshi Urano, Koichi Furukawa & Yota Murakami, "RNA Polymerase II Is Required for RNAi-Dependent Heterochromatin Assembly," *Science* 309 (2005): 467–469. Available online with registration (2011) at http://www.sciencemag.org/cgi/content/full/309/5733/467

48. Pavel Neumann, Huihuang Yan & Jiming Jiang, "The Centromeric Retrotransposons of Rice Are Transcribed and Differentially Processed by RNA Interference," *Genetics* 176 (2007): 749–761. Freely accessible (2011) at http://www.genetics.org/cgi/reprint/176/2/749

49. Angeline Eymery, Mary Callanan & Claire Vourc'h, "The secret message of heterochromatin: new insights into the mechanisms and function of centromeric and pericentric repeat sequence transcription," *International Journal of Developmental Biology* 53 (2009): 259–268. Freely accessible (2011) at http://www.ijdb.ehu.es/web/paper.php?doi=10.1387/ijdb.082673ae

50. Jonathan C. Lamb & James A. Birchler, "The role of DNA sequence in centromere formation," *Genome Biology* 4:5 (2003): 214. Freely accessible (2011) at http://genomebiology.com/content/pdf/gb-2003-4-5-214.pdf

51. Mary G. Schueler & Beth A. Sullivan, "Structural and functional dynamics of human centromeric chromatin," *Annual Review of Genomics and Human Genetics* 7 (2006): 301–313.

52. Beth A. Sullivan, Michael D. Blower & Gary H. Karpen, "Determining centromere identity: cyclical stories and forking paths," *Nature Reviews Genetics* 2 (2001): 584–596.

53. John J. Harrington, Gil Van Bokkelen, Robert W. Mays, Karen Gustashaw & Huntington F. Willard, "Formation of de novo centromeres and construction of first-generation human artificial microchromosomes," *Nature Genetics* 15 (1997): 345–355.

54. Desirée du Sart, Michael R. Cancilla, Elizabeth Earle, Jen-i Mao, Richard Saffery, Kellie M. Tainton, Paul Kalitsis, John Martyn, Alyssa E. Barry & K. H. Andy Choo, "A functional neo-centromere formed through activation of a latent human centromere and consisting of non-alpha-satellite DNA," *Nature Genetics* 16 (1997): 144–153.

55. Peter E. Warburton, "Chromosomal dynamics of human neocentromere formation," *Chromosome Research* 12 (2004): 617–626.

56. W. C. Earnshaw & N. Rothfield, "Identification of a family of human centromere

proteins using autoimmune sera from patients with scleroderma," *Chromosoma* 91 (1985): 313–321.

57. Douglas K. Palmer, Kathleen O'Day, Mark H. Wener, Brian S. Andrews & Robert L. Margolis, "A 17-kD Centromere Protein (CENP-A) Copurifies with Nucleosome Core Particles and with Histones," *Journal of Cell Biology* 104 (1987): 805–815. Freely accessible (2011) at http://jcb.rupress.org/content/104/4/805.long

58. Peter E. Warburton, Carol A. Cooke, Sylvie Bourassa, Omid Vafa, Beth A. Sullivan, Gail Stetten, Giorgio Gimelli, Dorothy Warburton, Chris Tyler-Smith, Kevin F. Sullivan, Guy G. Poirier & William C. Earnshaw, "Immunolocalization of CENP-A suggests a distinct nucleosome structure at the inner kinetochore plate of active centromeres," *Current Biology* 7 (1997): 901–904.

59. Aaron A. Van Hooser, Michael A. Mancini, C. David Allis, Kevin F. Sullivan & B. R. Brinkley, "The mammalian centromere: structural domains and the attenuation of chromatin modeling," *FASEB Journal* 13 Supplement (1999): S216-S220. Freely accessible (2011) at http://www.fasebj.org/cgi/reprint/13/9002/S216

60. Kinya Yoda, Satoshi Ando, Setsuo Morishita, Kenichi Houmura, Keiji Hashimoto, Kunio Takeyasu & Tuneko Okazaki, "Human centromere protein A (CENP-A) can replace histone H3 in nucleosome reconstitution in vitro," *Proceedings of the National Academy of Sciences USA* 97 (2000): 7266–7271. Freely accessible (2011) at http://www.pnas.org/content/97/13/7266.full.pdf+html

61. Ben E. Black, Melissa A. Brock, Sabrina Bédard, Virgil L. Woods, Jr. & Don W. Cleveland, "An epigenetic mark generated by the incorporation of CENP-A into centromeric nucleosomes," *Proceedings of the National Academy of Sciences USA* 104 (2007): 5008–5013. Freely accessible (2011) at http://www.pnas.org/content/104/12/5008.full.pdf+html

62. Mònica Torras-Llort, Olga Moreno-Moreno & Fernando Azorín, "Focus on the centre: the role of chromatin on the regulation of centromere identity and function," *EMBO Journal* 28 (2009): 2337–2348. Freely accessible (2011) at http://www.ncbi.nlm.nih.gov/pmc/articles/PMC2722248/pdf/emboj2009174a.pdf

63. Aaron A. Van Hooser, Ilia I. Ouspenski, Heather C. Gregson, Daniel A. Starr, Tim J. Yen, Michael L. Goldberg, Kyoko Yokomori, William C. Earnshaw, Kevin F. Sullivan & B. R. Brinkley, "Specification of kinetochore-forming chromatin by the histone H3 variant CENP-A," *Journal of Cell Science* 114 (2001): 3529–3542. Freely accessible (2011) at http://jcs.biologists.org/cgi/reprint/114/19/3529

64. Larissa J. Vos, Jakub K. Famulski & Gordon K.T. Chan, "How to build a centromere: from centromeric and pericentromeric chromatin to kinetochore assembly," *Biochemistry and Cell Biology* 84 (2006): 619–639. Freely accessible (2011) at http://article.pubs.nrc-cnrc.gc.ca/ppv/RPViewDoc?issn=0829-8211&volume=84&issue=4&startPage=619

65. Deborah L. Grady, Robert L. Ratliff, Donna L. Robinson, Erin C. McCanlies, Julianne Meyne & Robert K. Moyzis, "Highly conserved repetitive DNA sequences are present at human centromeres," *Proceedings of the National Academy of Sciences USA* 89 (1992): 1695–1699. Freely accessible (2011) at http://www.pnas.org/content/89/5/1695.full.pdf+html

66. Jiming Jiang, Shuhei Nasuda, Fenggao Dong, Christopher W. Sherrer, Sung-Sick Woo, Rod A. Wing, Bikram S. Gill & David C. Ward, "A conserved repetitive DNA element located in the centromeres of cereal chromosomes," *Proceedings of the National Academy of Sciences USA* 93 (1996): 14210–14213. Freely accessible

(2011) at http://www.pnas.org/content/93/24/14210.full.pdf+html

67. Huntington F. Willard, "Chromosome-Specific Organization of Human Alpha Satellite DNA," *American Journal of Human Genetics* 37 (1985): 524–532. Freely accessible (2011) at http://www.ncbi.nlm.nih.gov/pmc/articles/PMC1684601/pdf/ajhg00158-0092.pdf

68. John S. Waye & Huntington F. Willard, "Chromosome-specific alpha satellite DNA: nucleotide sequence analysis of the 2.0 kilobasepair repeat from the human X chromosome," *Nucleic Acids Research* 13 (1985): 2731–2743. Freely accessible (2011) at http://www.ncbi.nlm.nih.gov/pmc/articles/PMC341190/pdf/nar00302-0062.pdf

69. R. Heller, K. E. Brown, C. Burgtorf & W. R. A. Brown, "Mini-chromosomes derived from the human Y chromosome by telomere directed chromosome breakage," *Proceedings of the National Academy of Sciences USA* 93 (1996): 7125–7130. Freely accessible (2011) at http://www.pnas.org/content/93/14/7125.full.pdf+html

70. Terence D. Murphy and Gary H. Karpen, "Centromeres Take Flight: Alpha Satellite and the Quest for the Human Centromere," *Cell* 93 (1998): 317–320.

71. Brenda R. Grimes, Angela A. Rhoades & Huntington F. Willard, "Alpha-satellite DNA and vector composition influence rates of human artificial chromosome formation," *Molecular Therapy* 5 (2002): 798–805.

72. M. Katharine Rudd, Robert W. Mays, Stuart Schwartz & Huntington F. Willard, "Human Artificial Chromosomes with Alpha Satellite-Based De Novo Centromeres Show Increased Frequency of Nondisjunction and Anaphase Lag," *Molecular and Cellular Biology* 23 (2003): 7689–7697. Freely accessible (2011) at http://mcb.asm.org/cgi/reprint/23/21/7689

73. Gregory P. Copenhaver, Kathryn Nickel, Takashi Kuromori, Maria-Ines Benito, Samir Kaul, Xiaoying Lin, Michael Bevan, George Murphy, Barbara Harris, Laurence D. Parnell, W. Richard McCombie, Robert A. Martienssen, Marco Marra & Daphne Preuss, "Genetic Definition and Sequence Analysis of *Arabidopsis* Centromeres," *Science* 286 (1999): 2468–2474. Available online with registration (2011) at http://www.sciencemag.org/cgi/content/full/286/5449/2468

74. Wolfgang Haupt, Thilo C. Fischer, Sabine Winderl, Paul Fransz & Ramón A. Torres-Ruiz, "The centromere1 (CEN1) region of *Arabidopsis thaliana*: architecture and functional impact of chromatin," *The Plant Journal* 27 (2001): 285–296.

75. Sarah E. Hall, Gregory Kettler & Daphne Preuss, "Centromere Satellites from *Arabidopsis* Populations: Maintenance of Conserved and Variable Domains," *Genome Research* 13 (2003): 195–205. Freely accessible (2011) at http://genome.cshlp.org/content/13/2/195.full.pdf+html

76. Anthony W. I. Lo, Jeffrey M. Craig, Richard Saffery, Paul Kalitsis, Danielle V. Irvine, Elizabeth Earle, Dianna J. Magliano & K. H. Andy Choo, "A 330 kb CENP-A binding domain and altered replication timing at a human neocentromere," *EMBO Journal* 20 (2001): 2087–2096. Freely accessible (2011) at http://www.nature.com/emboj/journal/v20/n8/pdf/7593708a.pdf

77. Anderly C. Chueh, Lee H. Wong, Nicholas Wong & K.H. Andy Choo, "Variable and hierarchical size distribution of L1-retroelement-enriched CENP-A clusters within a functional human neocentromere," *Human Molecular Genetics* 14 (2005): 85–93. Freely accessible (2011) at http://hmg.oxfordjournals.org/cgi/reprint/14/1/85

78. Anderly C. Chueh, Emma L. Northrop, Kate H. Brettingham-Moore, K. H. Andy Choo & Lee H. Wong, "LINE Retrotransposon RNA Is an Essential

Structural and Functional Epigenetic Component of a Core Neocentromeric Chromatin," *PLoS Genetics* 5:1 (2009): e1000354. Freely accessible (2011) at http://www.plosgenetics.org/article/info%3Adoi%2F10.1371%2Fjournal.pgen.1000354

79. Stephen M. Stack, David B. Brown & William C. Dewey, "Visualization of interphase chromosomes," *Journal of Cell Science* 26 (1977): 281–299. Freely accessible (2011) at http://jcs.biologists.org/cgi/reprint/26/1/281

80. Jenny A. Croft, Joanna M. Bridger, Shelagh Boyle, Paul Perry, Peter Teague & Wendy A. Bickmore, "Differences in the Localization and Morphology of Chromosomes in the Human Nucleus," *Journal of Cell Biology* 145 (1999): 1119–1131. Freely accessible (2011) at http://jcb.rupress.org/content/145/6/1119.full.pdf+html

81. Heiner Albiez, Marion Cremer, Cinzia Tiberi, Lorella Vecchio, Lothar Schermelleh, Sandra Dittrich, Katrin Küpper, Boris Joffe, Tobias Thormeyer, Johann von Hase, Siwei Yang, Karl Rohr, Heinrich Leonhardt, Irina Solovei, Christoph Cremer, Stanislav Fakan & Thomas Cremer, "Chromatin domains and the interchromatin compartment form structurally defined and functionally interacting nuclear networks," *Chromosome Research* 14 (2006): 707–733.

82. Thomas Cremer & Marion Cremer, "Chromosome Territories," *Cold Spring Harbor Perspectives in Biology* 2 (2010): a003889. Freely accessible (2011) at http://cshperspectives.cshlp.org/content/2/3/a003889.full.pdf+html

83. Emanuela V. Volpi, Edith Chevret, Tania Jones, Radost Vatcheva, Jill Williamson, Stephan Beck, R. Duncan Campbell, Michelle Goldsworthy, Stephen H. Powis, Jiannis Ragoussis, John Trowsdale & Denise Sheer, "Large-scale chromatin organization of the major histocompatibility complex and other regions of human chromosome 6 and its response to interferon in interphase nuclei," *Journal of Cell Science* 113 (2000): 1565–1576. Freely accessible (2011) at http://jcs.biologists.org/cgi/reprint/113/9/1565

84. Li-Feng Zhang, Khanh D. Huynh & Jeannie T. Lee, "Perinucleolar targeting of the inactive X during S phase: evidence for a role in the maintenance of silencing," *Cell* 129 (2007): 693–706.

85. Christian Lanctôt, Thierry Cheutin, Marion Cremer, Giacomo Cavalli & Thomas Cremer, "Dynamic genome architecture in the nuclear space: regulation of gene expression in three dimensions," *Nature Reviews Genetics* 8 (2007): 104–115.

86. Boris Joffe, Heinrich Leonhardt & Irina Solovei, "Differentiation and large scale spatial organization of the genome," *Current Opinion in Genetics and Development* 20 (2010): 562–569.

87. M. R. Hübner & D. L. Spector, "Chromatin dynamics," *Annual Review of Biophysics* 39 (2010): 471–489.

88. Hideki Tanizawa, Osamu Iwasaki, Atsunari Tanaka, Joseph R. Capizzi, Priyankara Wickramasinghe, Mihee Lee, Zhiyan Fu & Ken-ichi Noma, "Mapping of long-range associations throughout the fission yeast genome reveals global genome organization linked to transcriptional regulation," *Nucleic Acids Research* 38 (2010): 8164–8177. Freely accessible (2011) at http://nar.oxfordjournals.org/content/38/22/8164.long

89. Luis A. Parada, Philip G. McQueen & Tom Misteli, "Tissue-specific spatial organization of genomes," *Genome Biology* 5:7 (2004): R44. Freely accessible (2011) at http://genomebiology.com/content/pdf/gb-2004-5-7-r44.pdf

90. Tom Sexton, Heiko Schober, Peter Fraser & Susan M. Gasser, "Gene regulation through nuclear organization," *Nature Structural & Molecular Biology* 14 (2007): 1049–1055.

91. Takumi Takizawa, Karen J. Meaburn & Tom Misteli, "The meaning of gene positioning," *Cell* 135 (2008): 9–13.

92. Erik D. Andrulis, Aaron M. Neiman, David C. Zappulla & Rolf Sternglanz, "Perinuclear localization of chromatin facilitates transcriptional silencing," *Nature* 394 (1998): 592–595.

93. Shelagh Boyle, Susan Gilchrist, Joanna M. Bridger, Nicola L. Mahy, Juliet A. Ellis & Wendy A. Bickmore, "The spatial organization of human chromosomes within the nuclei of normal and emerin-mutant cells," *Human Molecular Genetics* 10 (2001): 211–219. Freely accessible (2011) at http://hmg.oxfordjournals.org/content/10/3/211.full.pdf+html

94. Lars Guelen, Ludo Pagie, Emilie Brasset, Wouter Meuleman, Marius B. Faza, Wendy Talhout, Bert H. Eussen, Annelies de Klein, Lodewyk Wessels, Wouter de Laat & Bas van Steensel, "Domain organization of human chromosomes revealed by mapping of nuclear lamina interactions," *Nature* 453 (2008): 948–951.

95. Lee E. Finlan, Duncan Sproul, Inga Thomson, Shelagh Boyle, Elizabeth Kerr, Paul Perry, Bauke Ylstra, Jonathan R. Chubb & Wendy A. Bickmore, "Recruitment to the Nuclear Periphery Can Alter Expression of Genes in Human Cells," *PLoS Genetics* 4:3 (2008): e1000039. Freely accessible (2011) at http://www.plosgenetics.org/article/info%3Adoi%2F10.1371%2Fjournal.pgen.1000039

96. K. L. Reddy, J. M. Zullo, E. Bertolino & H. Singh, "Transcriptional repression mediated by repositioning of genes to the nuclear lamina," *Nature* 452 (2008): 243–247.

97. Myriam Ruault, Marion Dubarry & Angela Taddei, "Re-positioning genes to the nuclear envelope in mammalian cells: impact on transcription," *Trends in Genetics* 24 (2008): 574–581.

98. Davide Marenduzzo, Cristian Micheletti & Peter R. Cook, "Entropy-Driven Genome Organization," *Biophysical Journal* 90 (2006): 3712–3721. Freely accessible (2011) at http://www.ncbi.nlm.nih.gov/pmc/articles/PMC1440752/pdf/3712.pdf

99. Peter R. Cook & Davide Marenduzzo, "Entropic organization of interphase chromosomes," *Journal of Cell Biology* 186 (2009): 825–834. Freely accessible (2011) at http://jcb.rupress.org/content/186/6/825.full.pdf+html

100. Jan Postberg, Olga Alexandrova, Thomas Cremer & Hans J. Lipps, "Exploiting nuclear duality of ciliates to analyse topological requirements for DNA replication and transcription," *Journal of Cell Science* 118 (2005): 3973–3983. Freely accessible (2011) at http://jcs.biologists.org/cgi/reprint/118/17/3973

101. L. D. Carter-Dawson & M. M. LaVail, "Rods and cones in the mouse retina. I. Structural analysis using light and electron microscopy," *Journal of Comparative Neurology* 188 (1979): 245–262.

102. Seth Blackshaw, Rebecca E. Fraioli, Takahisa Furukawa & Constance L. Cepko, "Comprehensive analysis of photoreceptor gene expression and the identification of candidate retinal disease genes," *Cell* 107 (2001): 579–589.

103. Dominique Helmlinger, Sara Hardy, Gretta Abou-Sleymane, Adrien Eberlin, Aaron B. Bowman, Anne Gansmüller, Serge Picaud, Huda Y. Zoghbi, Yvon Trottier, Làszlò Tora & Didier Devys, "Glutamine-Expanded Ataxin-7 Alters TFTC/STAGA Recruitment and Chromatin Structure Leading to Photoreceptor Dysfunction," *PLoS Biology* 4:3 (2006): e67. Freely accessible (2011) at http://www.plosbiology.org/article/info%3Adoi%2F10.1371%2Fjournal.pbio.0040067

104. Irina Solovei, Moritz Kreysing, Christian Lanctôt, Süleyman Kösem, Leo Peichl, Thomas Cremer, Jochen Guck &

Boris Joffe, "Nuclear Architecture of Rod Photoreceptor Cells Adapts to Vision in Mammalian Evolution," *Cell* 137 (2009): 356–368.

105. Irina Solovei, Moritz Kreysing, Christian Lanctôt, Süleyman Kösem, Leo Peichl, Thomas Cremer, Jochen Guck & Boris Joffe, "Nuclear Architecture of Rod Photoreceptor Cells Adapts to Vision in Mammalian Evolution," *Cell* 137 (2009): Supplementary Data. Freely accessible (2011) at http://download. cell.com/mmcs/journals/0092-8674/PIIS0092867409001378.mmc1.pdf

106. Caroline Kizilyaprak, Danièle Spehner, Didier Devys & Patrick Schultz, "*In Vivo* Chromatin Organization of Mouse Rod Photoreceptors Correlates with Histone Modifications," *PLoS One* 5:6 (2010): e11039. Freely accessible (2011) at http://www.plosone.org/article/info%3Adoi%2F10.1371%2Fjournal.pone.0011039

107. Moritz Kreysing, Lars Boyde, Jochen Guck & Kevin J. Chalut, "Physical insight into light scattering by photoreceptor cell nuclei," *Optics Letters* 35 (2010): 2639–2641.

8. Some Recent Defenders of Junk DNA

1. Shinji Hirotsune, Noriyuki Yoshida, Amy Chen, Lisa Garrett, Fumihiro Sugiyama, Satoru Takahashi, Ken-ichi Yagami, Anthony Wynshaw-Boris & Atsushi Yoshiki, "An expressed pseudogene regulates the messenger-RNA stability of its homologous coding gene," *Nature* 423 (2003): 91–96.

2. William S. Harris & John H. Calvert, "Intelligent Design: The Scientific Alternative to Evolution," *The National Catholic Bioethics Quarterly* (August 2003): 531–561. Freely accessible (2011) at http://www.intelligentdesignnetwork.org/NCBQ3_3HarrisCalvert.pdf

3. Todd A. Gray, Alison Wilson, Patrick J. Fortin & Robert D. Nicholls, "The putatively functional *Mkrn1-p1* pseudogene is neither expressed nor imprinted, nor does it regulate its source gene in trans," *Proceedings of the National Academy of Sciences USA* 103 (2006): 12039–12044. Freely accessible (2011) at http://www.pnas.org/content/103/32/12039.full.pdf+html

4. José Manuel Franco-Zorrilla, Adrián Valli, Marco Todesco, Isabel Mateos, María Isabel Puga, Ignacio Rubio-Somoza, Antonio Leyva, Detlef Weigel, Juan Antonio García & Javier Paz-Ares, "Target mimicry provides a new mechanism for regulation of microRNA activity," *Nature Genetics* 39 (2007): 1033–1037.

5. Laura Poliseno, Leonardo Salmena, Jiangwen Zhang, Brett Carver, William J. Haveman & Pier Paolo Pandolfi, "A coding-independent function of gene and pseudogene mRNAs regulates tumour biology," *Nature* 465 (2010): 1033–1038.

6. Kayoko Yamada, Jun Lim, Joseph M. Dale, Huaming Chen, Paul Shinn, Curtis J. Palm, Audrey M. Southwick, Hank C. Wu, Christopher Kim, Michelle Nguyen, Paul Pham, Rosa Cheuk, George Karlin-Newmann, Shirley X. Liu, Bao Lam, Hitomi Sakano, Troy Wu, Guixia Yu, Molly Miranda, Hong L. Quach, Matthew Tripp, Charlie H. Chang, Jeong M. Lee, Mitsue Toriumi, Marie M. H. Chan, Carolyn C. Tang, Courtney S. Onodera, Justine M. Deng, Kenji Akiyama, Yasser Ansari, Takahiro Arakawa, Jenny Banh, Fumika Banno, Leah Bowser, Shelise Brooks, Piero Carninci, Qimin Chao, Nathan Choy, Akiko Enju, Andrew D. Goldsmith, Mani Gurjal, Nancy F. Hansen, Yoshihide Hayashizaki, Chanda Johnson-Hopson, Vickie W. Hsuan, Kei Iida, Meagan Karnes, Shehnaz Khan, Eric Koesema, Junko Ishida, Paul X. Jiang, Ted Jones, Jun Kawai, Asako Kamiya, Cristina Meyers, Maiko Nakajima, Mari Narusaka, Motoaki Seki, Tetsuya Sakurai, Masakazu Satou, Racquel Tamse, Maria Vaysberg, Erika

K. Wallender, Cecilia Wong, Yuki Yamamura, Shiaulou Yuan, Kazuo Shinozaki, Ronald W. Davis, Athanasios Theologis & Joseph R. Ecker, "Empirical Analysis of Transcriptional Activity in the *Arabidopsis* Genome," *Science* 302 (2003): 842–846. Available online with registration (2011) at http://www.sciencemag.org/cgi/content/full/302/5646/842

7. Jason M. Johnson, Stephen Edwards, Daniel Shoemaker & Eric E. Schadt, "Dark matter in the genome: evidence of widespread transcription detected by microarray tiling experiments," *Trends in Genetics* 21 (2005): 93–102.

8. Guy Riddihough, "In the Forests of RNA Dark Matter," *Science* 309 (2005): 1507.

9. Harm van Bakel & Timothy R. Hughes, "Establishing legitimacy and function in the new transcriptome," *Briefings in Functional Genomics & Proteomics* 8 (2009): 424–436.

10. Harm van Bakel, Corey Nislow, Benjamin J. Blencowe & Timothy R. Hughes, "Most 'Dark Matter' Transcripts Are Associated With Known Genes," *PLoS Biology* 8:5 (2010): e1000371. Freely accessible (2011) at http://www.plosbiology.org/article/info%3Adoi%2F10.1371%2Fjournal.pbio.1000371

11. The ENCODE Project Consortium, "Identification and analysis of functional elements in 1% of the human genome by the ENCODE pilot project," *Nature* 447 (2007): 799–816. Freely accessible (2011) at http://www.ncbi.nlm.nih.gov/pmc/articles/PMC2212820/pdf/nihms27513.pdf

12. Richard Robinson, "Dark Matter Transcripts: Sound and Fury, Signifying Nothing?" *PLoS Biology* 8:5 (2010): e1000370. Freely accessible (2011) at http://www.plosbiology.org/article/info%3Adoi%2F10.1371%2Fjournal.pbio.1000370

13. Carl Zimmer, "How Many Sparks in the Genome?" *Discover Magazine: The Loom* (May 19, 2010). Freely accessible (2011) at http://blogs.discovermagazine.com/loom/2010/05/19/how-many-sparks-in-the-genome/

14. RepeatMasker, *Institute for Systems Biology*. Freely accessible (2011) at http://www.repeatmasker.org/

15. Jill Cheng, Philipp Kapranov, Jorg Drenkow, Sujit Dike, Shane Brubaker, Sandeep Patel, Jeffrey Long, David Stern, Hari Tammana, Gregg Helt, Victor Sementchenko, Antonio Piccolboni, Stefan Bekiranov, Dione K. Bailey, Madhavan Ganesh, Srinka Ghosh, Ian Bell, Daniela S. Gerhard & Thomas R. Gingeras, "Transcriptional Maps of 10 Human Chromosomes at 5-Nucleotide Resolution," *Science* 308 (2005): 1149–1154.

16. P. Z. Myers, "Junk DNA is still junk," *The Panda's Thumb* (May 19, 2010). Freely accessible (2011) at http://pandasthumb.org/archives/2010/05/junk-dna-is-sti.html

17. Philipp Kapranov, Georges St. Laurent, Tal Raz, Fatih Ozsolak, C. Patrick Reynolds, Poul H. B. Sorensen, Gregory Reaman, Patrice Milos, Robert J. Arceci, John F. Thompson & Timothy J. Triche, "The majority of total nuclear-encoded non-ribosomal RNA in a human cell is 'dark matter' un-annotated RNA," *BMC Biology* 8:1 (2010): 149. Freely accessible (2011) at http://www.biomedcentral.com/1741-7007/8/149

18. Mark Blaxter, "Revealing the Dark Matter of the Genome," *Science* 330 (2010): 1758–1759.

19. Philipp Kapranov, Aarron T. Willingham & Thomas R. Gingeras, "Genome-wide transcription and the implications for genomic organization," *Nature Reviews Genetics* 8 (2007): 413–423.

20. John L. Rinn, Michael Kertesz, Jordon K. Wang, Sharon L. Squazzo, Xiao Xu, Samantha A. Brugmann, Henry Goodnough, Jill A. Helms, Peggy J. Farnham, Eran Segal, and Howard Y. Chang, "Functional

Demarcation of Active and Silent Chromatin Domains in Human HOX Loci by Non-Coding RNAs," *Cell* 129 (2007): 1311–1323. Freely accessible (2011) at http://www.ncbi.nlm.nih.gov/pmc/articles/PMC2084369/?tool=pubmed

21. Elizabeth Pennisi, "Shining a Light on the Genome's 'Dark Matter,'" *Science* 330 (2010): 1614.

22. T. Ryan Gregory, "Junk DNA: let me say it one more time," *Genomicron* (September 16, 2007). Freely accessible (2011) at http://www.genomicron.evolverzone.com/2007/09/junk-dna-let-me-say-it-one-more-time/

23. T. Ryan Gregory, "The onion test," *Genomicron* (April 25, 2007). Freely accessible (2011) at http://www.genomicron.evolverzone.com/2007/04/onion-test/

24. Roger Vendrely & Colette Vendrely, "La teneur du noyau cellulaire en acide désoxyribonucléique à travers les organes, les individus et les espèces animales: Techniques et premiers resultants," *Experientia* 4 (1948): 434–436.

25. A. E. Mirsky & Hans Ris, "The Desoxyribonucleic Acid Content of Animal Cells and Its Evolutionary Significance," *Journal of General Physiology* 34 (1951): 451–462. Freely accessible (2011) at http://www.ncbi.nlm.nih.gov/pmc/articles/PMC2147229/pdf/451.pdf

26. C. A. Thomas, Jr., "The Genetic Organization of Chromosomes," *Annual Review of Genetics* 5 (1971): 237–256.

27. Joseph G. Gall, "Chromosome Structure and the C-Value Paradox," *Journal of Cell Biology* 91 (1981): 3s-14s. Freely accessible (2011) at http://jcb.rupress.org/content/91/3/3s.full.pdf

28. Gordon P. Moore, "The C-Value Paradox," *BioScience* 34 (July/August 1984): 425–429.

29. Wen-Hsiung Li, *Molecular Evolution* (Sunderland, MA: Sinauer Associates, 1997), pp. 379–384.

30. Daniel L. Hartl, "Molecular melodies in high and low C," *Nature Reviews Genetics* 1 (2000): 145–149.

31. Thomas Cavalier-Smith, "Nuclear volume control by nucleoskeletal DNA, selection for cell volume and cell growth rate, and the solution of the DNA C-value paradox," *Journal of Cell Science* 34 (1978): 247–278. Freely accessible (2011) at http://jcs.biologists.org/cgi/reprint/34/1/247

32. Thomas Cavalier-Smith, "Cell Volume and the Evolution of Eukaryotic Genome Size," pp. 105–184 in Thomas Cavalier-Smith (editor), *The Evolution of Genome Size* (Chichester, UK: John Wiley & Sons, 1985).

33. Alexander E. Vinogradov, "Nucleotypic Effect in Homeotherms: Body-Mass-Corrected Basal Metabolic Rate of Mammals Is Related to Genome Size," *Evolution* 49 (1995): 1249–1259.

34. R. A. Van Den Bussche, J. L. Longmire & R. J. Baker, "How bats achieve a small C-value: frequency of repetitive DNA in *Macrotus*," *Mammalian Genome* 6 (1995): 521–525.

35. Austin L. Hughes & Marianne K. Hughes, "Small genomes for better flyers," *Nature* 377 (1995): 391.

36. Alexander E. Vinogradov, "Nucleotypic Effect in Homeotherms: Body-Mass Independent Resting Metabolic Rate of Passerine Birds Is Related to Genome Size," *Evolution* 51 (1997): 220–225.

37. T. Ryan Gregory, "A bird's-eye view of the C-value enigma: genome size, cell size, and metabolic rate in the class aves," *Evolution* 56 (2002): 121–130.

38. Stanley K. Sessions & Allan Larson, "Developmental Correlates of Genome Size in Plethodontid Salamanders and their Implications for Genome Evolution," *Evolution* 41 (1987): 1239–1251.

39. T. Ryan Gregory & Paul D. N. Hebert, "The Modulation of DNA Content: Proximate Causes and Ultimate Consequences," *Genome Research* 9 (1999): 317–324. Freely

accessible (2011) at http://genome.cshlp.org/content/9/4/317.full.pdf+html

40. T. Ryan Gregory, "Genome size and developmental complexity," *Genetica* 115 (2002): 131–146.

41. T. Ryan Gregory, "The C-value Enigma in Plants and Animals: A Review of Parallels and an Appeal for Partnership," *Annals of Botany* 95 (2005): 133–146. Freely accessible (2011) at http://aob.oxfordjournals.org/content/95/1/133.full.pdf+html

42. T. Ryan Gregory & J. S. Johnston, "Genome size diversity in the family Drosophilidae," *Heredity* 101 (2008): 228–238.

43. Emile Zuckerkandl, "Gene control in eukaryotes and the c-value paradox 'excess' DNA as an impediment to transcription of coding sequences," *Journal of Molecular Evolution* 9 (1976): 73–104.

44. Sean Luke, "Evolutionary computation and the c-value paradox," *Proceedings of the 2005 conference on Genetic and Evolutionary Computation* (2005): 91–97.

45. Thomas Cavalier-Smith, "Economy, speed and size matter: evolutionary forces driving nuclear genome miniaturization and expansion," *Annals of Botany* 95 (2005): 147–175.

46. Ryan J. Taft, Michael Pheasant & John S. Mattick, "The relationship between non-protein-coding DNA and eukaryotic complexity," *BioEssays* 29 (2007): 288–299.

47. L. I. Patrushev & I. G. Minkevich, "The Problem of the Eukaryotic Genome Size," *Biochemistry* (Moscow) 73 (2008): 1519–1552. Freely accessible (2011) at http://protein.bio.msu.ru/biokhimiya/contents/v73/full/73131519.html

48. Eduard Kejnovsky, Ilia J. Leitch & Andrew R. Leitch, "Contrasting evolutionary dynamics between angiosperm and mammalian genomes," *Trends in Ecology and Evolution* 24 (2009): 572–582.

49. T. Ryan Gregory, "Coincidence, coevolution, or causation? DNA content, cell size, and the C-value enigma," *Biological Reviews*

of the Cambridge Philosophical Society 76 (2001): 65–101.

50. T. Ryan Gregory, "Genome Size Evolution in Animals," pp. 3–87 in T. Ryan Gregory (editor), *The Evolution of the Genome* (Amsterdam: Elsevier, 2005), pp. 48–49.

51. T. Ryan Gregory, "An opportunity for ID to be scientific," *Genomicron* (July 10, 2007). Freely accessible (2011) at http://www.genomicron.evolverzone.com/2007/07/opportunity-for-id-to-be-scientific/

52. William A. Dembski, *Intelligent Design: The Bridge Between Science and Theology* (Downer's Grove, IL: InterVarsity Press, 2002), p. 150.

53. William A. Dembski, *The Design Revolution* (Downer's Grove, IL: InterVarsity Press, 2004), p. 272.

54. T. Ryan Gregory, "Function, non-function, some function: a brief history of junk DNA," *Genomicron* (June 14, 2007). Freely accessible (2011) at http://www.genomicron.evolverzone.com/2007/06/function-non-function-some-function/

9. A SUMMARY OF THE CASE FOR FUNCTIONALITY IN JUNK DNA

1. Luis M. Mendes Soares & Juan Valcárcel, "The expanding transcriptome: the genome as the 'Book of Sand'," *EMBO Journal* 25 (2006): 923–931. Available online with registration (2011) at http://www.nature.com/emboj/journal/v25/n5/full/7601023a.html

2. Piero Carninci, Jun Yasuda & Yoshihide Hayashizaki, "Multifaceted mammalian transcriptome," *Current Opinion in Cell Biology* 20 (2008): 274–280.

3. Yoseph Barash, John A. Calarco, Weijun Gao, Qun Pan, Xinchen Wang, Ofer Shai, Benjamin J. Blencowe & Brendan J. Frey, "Deciphering the splicing code," *Nature* 465 (2010): 53–59.

4. Laura Poliseno, Leonardo Salmena, Jiangwen Zhang, Brett Carver, William

J. Haveman & Pier Paolo Pandolfi, "A coding-independent function of gene and pseudogene mRNAs regulates tumour biology," *Nature* 465 (2010): 1033–1038.

5. Ryan D. Walters, Jennifer F. Kugel & James A. Goodrich, "InvAluable junk: the cellular impact and function of *Alu* and B2 RNAs," *IUBMB Life* 61 (2009): 831–837.

6. Victoria V. Lunyak, Gratien G. Prefontaine, Esperanza Núñez, Thorsten Cramer, Bong-Gun Ju, Kenneth A. Ohgi, Kasey Hutt, Rosa Roy, Angel García-Díaz, Xiaoyan Zhu, Yun Yung, Lluís Montoliu, Christopher K. Glass & Michael G. Rosenfeld, "Developmentally regulated activation of a SINE B2 repeat as a domain boundary in organogenesis," *Science* 317 (2007): 248–251.

7. Jonathan P. Stoye, "Proviral protein provides placental function," *Proceedings of the National Academy of Sciences USA* 106 (2009): 11827–11828. Freely accessible (2011) at http://www.pnas.org/content/106/29/11827.full.pdf+html

8. Emile Zuckerkandl, "Why so many non-coding nucleotides? The eukaryote genome as an epigenetic machine," *Genetica* 115 (2002): 105–129.

9. Stephen C. J. Parker, Loren Hansen, Hatice Ozel Abaan, Thomas D. Tullius & Elliott H. Margulies, "Local DNA Topography Correlates with Functional Noncoding Regions of the Human Genome," *Science* 324 (2009): 389–392.

10. T. Ryan Gregory, "Junk DNA: let me say it one more time," *Genomicron* (September 16, 2007). Freely accessible (2011) at http://www.genomicron.evolverzone.com/2007/09/junk-dna-let-me-say-it-one-more-time/

11. T. Ryan Gregory, "The onion test," *Genomicron* (April 25, 2007). Freely accessible (2011) at http://www.genomicron.evolverzone.com/2007/04/onion-test/

12. Harm van Bakel, Corey Nislow, Benjamin J. Blencowe & Timothy R. Hughes, "Most 'Dark Matter' Transcripts Are Associated With Known Genes," *PLoS Biology* 8 (2010): e1000371. Freely accessible (2011) at http://www.plosbiology.org/article/info%3Adoi%2F10.1371%2Fjournal.pbio.1000371

13. P. Z. Myers, "Junk DNA is still junk," *The Panda's Thumb* (May 19, 2010). Freely accessible (2011) at http://pandasthumb.org/archives/2010/05/junk-dna-is-sti.html

14. Philipp Kapranov, Georges St. Laurent, Tal Raz, Fatih Ozsolak, C. Patrick Reynolds, Poul H. B. Sorensen, Gregory Reaman, Patrice Milos, Robert J. Arceci, John F. Thompson & Timothy J. Triche, "The majority of total nuclear-encoded non-ribosomal RNA in a human cell is 'dark matter' un-annotated RNA," *BMC Biology* 8:1 (2010): 149. Freely accessible (2011) at http://www.biomedcentral.com/1741-7007/8/149

10. From Junk DNA to a New Understanding of the Genome

1. Francis S. Collins, *The Language of God: A Scientist Presents Evidence for Belief* (New York: Free Press, 2006), p. 136.

2. John C. Avise, *Inside the Human Genome: A Case for Non-Intelligent Design* (Oxford: Oxford University Press, 2010), p. 115.

3. Douglas J. Futuyma, *Evolution* (Sunderland, MA: Sinauer Associates, 2005), p. 456.

4. PubMed. Freely accessible (2011) at http://www.ncbi.nlm.nih.gov/sites/pubmed

5. James A. Shapiro & Richard von Sternberg, "Why repetitive DNA is essential to genome function," *Biological Reviews* 80 (2005): 227–250. Freely accessible (2011) at http://shapiro.bsd.uchicago.edu/index3.html?content=genome.html

6. The ENCODE Project Consortium, "Identification and analysis of functional elements in 1% of the human genome by the ENCODE pilot project," *Nature*

447 (2007): 799–816. Freely accessible (2011) at http://www.ncbi.nlm.nih.gov/pmc/articles/PMC2212820/pdf/nihms27513.pdf

7. Geoff Spencer & Anna-Lynn Wegener, "New Findings Challenge Established Views on Human Genome," *NIH News* (June 13, 2007). Freely accessible (2011) at http://genome.gov/25521554

8. Catherine Shaffer, "One Scientist's Junk Is a Creationist's Treasure," *Wired Magazine Blog* (June 13, 2007). Freely accessible (2011) at http://www.wired.com/science/discoveries/news/2007/06/junk_dna

9. Francis S. Collins, *The Language of Life: DNA and the Revolution in Personalized Medicine* (New York: HarperCollins, 2010), pp. 5–6, 9–10.

10. Collins, *The Language of Life*, p. 293.

11. Francis S. Collins, *The Language of God: A Scientist Presents Evidence for Belief* (New York: Free Press, 2006), pp. 199–203.

12. "About the BioLogos Foundation," BioLogos. Freely accessible (2011) at http://biologos.org/about

13. Darrel Falk, "A Rejoinder to Part II of Stephen C. Meyer's Response to Francisco Ayala," BioLogos Forum (March 11, 2010). Freely accessible (2011) at http://biologos.org/blog/a-rejoinder-to-meyer-2

14. Karl Giberson, "Does Intelligent Design Really Explain a Complex and Puzzling World?" BioLogos Forum (March 15, 2010). Freely accessible (2011) at http://biologos.org/blog/does-intelligent-design-really-explain-a-complex-and-puzzling-world/

15. Center for Science & Culture Staff, "Bibliography of Supplementary Resources for Science Instruction," *Discovery Institute* (March 11, 2002). Freely accessible (2011) at http://www.discovery.org/a/1127

16. NCSE Staff, "Analysis of the Discovery Institute's Bibliography," *National Center for Science Education* (June 1, 2002). Freely accessible (2011) at http://ncse.com/creationism/general/analysis-discovery-institutes-bibliography

17. NCSE Staff, "Analysis of the Discovery Institute's Bibliography: Appendix," *National Center for Science Education* (June 1, 2002). Freely accessible (2011) at http://ncse.com/creationism/general/analysis-discovery-institutes-bibliography-appendix

18. Center for Science & Culture Staff, "Questions and Answers About the Discovery Institute's Bibliography of Supplementary Resources for Ohio Science Instruction," *Discovery Institute* (April 15, 2002). Freely accessible (2011) at http://www.arn.org/docs2/news/discoveryresponse-toncse041702.htm

19. Charles Darwin, *The Origin of Species by Means of Natural Selection* (London: John Murray, 1859, p. 459. Freely accessible (2011) at http://darwin-online.org.uk/content/frameset?viewtype=side&itemID=F373&pageseq=477

20. Darwin, *The Origin of Species*, p. 437. Freely accessible (2011) at http://darwin-online.org.uk/content/frameset?viewtype=side&itemID=F373&pageseq=455

21. Jonathan Wells, "Darwin's Straw God Argument," *Journal of Interdisciplinary Studies* 22 (2010): 67–88. An earlier version (December 31, 2008) is freely accessible (2011) at http://www.discovery.org/a/8101

22. Neal C. Gillespie, *Charles Darwin and the Problem of Creation* (Chicago: The University of Chicago Press, 1979), pp. 124–125, 146.

23. Cornelius G. Hunter, *Darwin's God* (Grand Rapids, MI: Brazos Press, 2001), pp. 48–49, 84, 158.

24. Paul A. Nelson, "The role of theology in current evolutionary reasoning," *Biology and Philosophy* 11 (1996): 493 – 517.

25. Gregory Radick, "Deviance, Darwinina-Style," *Metascience* (2005) 14:453–457. Freely accessible (2011) at http://www.

personal.leeds.ac.uk/~phlgmr/Greg%20
Articles/Deviance.pdf

26. Abigail J. Lustig, "Natural Atheology,"
pp. 69–83 in Abigail J. Lustig, Robert J.
Richards & Michael Ruse (editors), *Darwinian Heresies* (Cambridge: Cambridge
University Press: 2004).

27. Elliott Sober, *Evidence and Evolution*
(Cambridge: Cambridge University Press,
2008), pp. 126–128.

28. Steven R. Scadding, "Vestigial organs
do not provide scientific evidence for
evolution," *Evolutionary Theory* 6 (1982):
171–173.

29. Jonathan Wells, "Darwin of the Gaps:
Francis Collins's Premature Surrender," pp.
117–128 in Jay W. Richards (editor), *God
and Evolution* (Seattle, WA: Discovery
Institute Press, 2010). An earlier version
(March 26, 2008) is freely accessible (2011)
at http://www.discovery.org/a/4529

30. Miller, "Life's Grand Design" (1994).
Freely accessible (2011) at http://www.
millerandlevine.com/km/evol/lgd/
index.html

31. Bill Gates, *The Road Ahead* (New York:
Penguin Books, 1995), p. 188.

32. Stephen C. Meyer, "The origin of biological information and the higher taxonomic
categories," *Proceedings of the Biological
Society of Washington* 117 (2004): 213–239.
Freely accessible (2011) at http://www.
discovery.org/a/2177

33. Stephen C. Meyer, *Signature in the Cell:
DNA and the Evidence for Intelligent Design*
(New York: HarperCollins, 2009). More
information available online (2011) at
http://www.discovery.org/a/12311

34. William A. Dembski, *The Design Revolution: Answering the Toughest Questions
about Intelligent Design* (Downer's Grove,
IL: InterVarsity Press, 2004), pp. 317.

35. Leon Brillouin, *Science and Information
Theory*, Second Edition (New York: Academic Press, 1956).

36. Hubert P. Yockey, *Information Theory
and Molecular Biology* (Cambridge: Cambridge University Press, 1992).

37. William A. Dembski, *The Design Inference* (Cambridge: Cambridge University
Press, 1998).

38. Wen-Yu Chung, Samir Wadhawan,
Radek Szklarczyk, Sergei Kosakovsky
Pond & Anton Nekrutenko, "A first
look at ARFome: dual-coding genes
in mammalian genomes," *PLoS Computational Biology* 3:5 (2007): e91.
Freely Accessible (2011) at http://
www.ploscompbiol.org/article/
info%3Adoi%2F10.1371%2Fjournal.
pcbi.0030091

39. Shalev Itzkovitz & Uri Alon, "The genetic code is nearly optimal for allowing additional information within protein-coding
sequences," *Genome Research* 17 (2007):
405–412. Freely accessible (2011) at http://
genome.cshlp.org/content/17/4/405.full.
pdf+html

40. Tobias Bollenbach, Kalin Vetsigian &
Roy Kishony, "Evolution and multilevel
optimization of the genetic code," *Genome
Research* 17 (2007): 401–404. Freely accessible (2011) at http://genome.cshlp.org/
content/17/4/401.full.pdf+html

41. Dembski, *The Design Revolution*, pp. 317.

42. Danielle S. Bassett, Daniel L. Greenfield,
Andreas Meyer-Lindenberg, Daniel R.
Weinberger, Simon W. Moore & Edward
T. Bullmore, "Efficient physical embedding
of topologically complex information processing networks in brains and computer
circuits," *PLoS Computational Biology* 6:4
(2010): e1000748. Freely accessible (2011)
at http://www.ploscompbiol.org/article/
info%3Adoi%2F10.1371%2Fjournal.
pcbi.1000748

43. Richard v. Sternberg, "On the Roles of
Repetitive DNA Elements in the Context
of a Unified Genomic–Epigenetic System,"
Annals of the New York Academy of Sciences
981 (2002): 154–188.

44. Richard v. Sternberg, "DNA Codes and Information: Formal Structures and Relational Causes," *Acta Biotheoretica* 56 (2008): 205–232.

45. Richard v. Sternberg & James A. Shapiro, "How repeated retroelements format genome function," *Cytogenetic and Genome Research* 110 (2005): 108–116.

APPENDIX. THE VITAMIN C PSEUDOGENE

1. Juan M. Navia & Charles E. Hunt, "Nutrition, Nutritional Diseases, and Nutrition Research Applications," pp. 235–267 in Joseph E. Wagner & Patrick J. Manning (editors), *The Biology of the Guinea Pig* (New York: Academic Press, 1976).

2. Yasuo Nakajima, Totada R. Shantha & Geoffrey H. Bourne, "Histochemical detection of L-gulonolactone: phenazine methosulfate oxidoreductase activity in several mammals with special reference to synthesis of vitamin C in primates," *Histochemie* 18 (1969): 293–301.

3. R. N. Roy & B. C. Guha, "Species Differences in regard to the Biosynthesis of Ascorbic Acid," *Nature* 182 (1958): 319–320.

4. Elmer C. Birney, Robert Jenness & Kathleen M. Ayaz, "Inability of bats to synthesise L-ascorbic acid," *Nature* 260 (1976): 626–628.

5. Jie Cui, Yi-Hsuan Pan, Yijian Zhang, Gareth Jones & Shuyi Zhang, "Progressive Pseudogenization: Vitamin C synthesis and Its Loss in Bats," *Molecular Biology and Evolution* (October 29, 2010). doi:10.1093/molbev/msq286.

6. C. Ray Chaudhuri & I. B. Chatterjee, "L-Ascorbic Acid Synthesis in Birds: Phylogenetic Trend," *Science* 164 (1969): 435–436.

7. Carlos Martínez del Rio, "Can Passerines Synthesize Vitamin C?" *The Auk* 114 (1997): 513–516.

8. J. E. Halver, R. R. Smith, B. M. Tolbert & E. M. Baker, "Utilization of Ascorbic Acid in Fish," *Annals of the New York Academy of Sciences* 258 (1975): 81–102.

9. Régis Moreau & Konrad Dabrowski, "Gulonolactone oxidase presence in fishes: activity and significance," pp. 14–32 in Konrad Dabrowski (editor), *Ascorbic Acid in Aquatic Organisms* (Boca Raton, LA: CRC Press, 2001).

10. Morimitsu Nishikimi & Kunio Yagi, "Molecular basis for the deficiency in humans of gulonolactone oxidase, a key enzyme for ascorbic acid biosynthesis," *American Journal of Clinical Nutrition* 54 (1991): 1203S-1208S. Freely accessible (2011) at http://www.ajcn.org/cgi/reprint/54/6/1203S

11. Morimitsu Nishikimi, Ryuichi Fukuyama, Sinsei Minoshima, Nobuyoshi Shimizu & Kunio Yagi, "Cloning and Chromosomal Mapping of the Human Nonfunctional Gene for L-Gulono-gamma-lactone Oxidase, the Enzyme for L-Ascorbic Acid Biosynthesis Missing in Man," *Journal of Biological Chemistry* 269 (1994): 13685–13688. Freely accessible (2011) at http://www.jbc.org/content/269/18/13685.long

12. Kenneth R. Miller, *Only a Theory: Evolution and the Battle for America's Soul* (New York: Viking, 2008), p. 98.

13. Miller, *Only a Theory*, p. 99.

14. Jonathan Wells, "Should We Stop Criticizing the Doctrine of Universal Common Ancestry?" *Access Research Network* (November 3, 2001). Freely accessible (2011) at http://www.arn.org/docs/wells/jw_criticizingcommonancestry1103.htm

15. William A. Dembski, *The Design Revolution: Answering the Toughest Questions About Intelligent Design* (Downer's Grove, IL: InterVarsity Press, 2004), p. 42.

16. Stephen C. Meyer, "Intelligent Design vs. Evolution," *Think Tank With Ben Wattenberg* (October 12, 2006). Freely accessible (2011) at http://www.pbs.org/thinktank/transcript1244.html

17. Paul A. Nelson, "Design and Common Ancestry," *Evolution News & Views* (May 7, 2007). Freely accessible (2011) at http://www.evolutionnews.org/2007/05/_most_people_including.html

18. Casey Luskin, "Wikipedia 'Intelligent Design' Entry Selectively Cites Poll Data to Present Misleading Picture of Support for Intelligent Design," *Evolution News & Views* (May 8, 2007). Freely accessible (2011) at http://www.evolutionnews.org/2007/05/wikipedia_intelligent_design_e003542.html

19. Michael J. Behe, *Darwin's Black Box: The Biochemical Challenge to Evolution* (New York: The Free Press, 1996), p. 231.

20. W. Ford Doolittle, "The practice of classification and the theory of evolution, and what the demise of Charles Darwin's tree of life hypothesis means for both of them," *Philosophical Transactions of the Royal Society of London B* 364 (2009): 2221–2228.

21. Carl R. Woese & Nigel Goldenfeld, "How the Microbial World Saved Evolution from the Scylla of Molecular Biology and the Charybdis of the Modern Synthesis," *Microbiology and Molecular Biology Reviews* 73 (2009): 14–21. Freely accessible (2011) at http://mmbr.asm.org/cgi/reprint/73/1/14

22. David G. Popovich & Ellen S. Dierenfeld, "Gorilla Nutrition," in J. Ogden & D. Wharton (editors), *Management of Gorillas in Captivity* (Silver Spring, MD: American Association of Zoos and Aquariums, 1997). Freely accessible (2011) at http://www.nagonline.net/HUSBANDRY/Diets%20pdf/Gorilla%20Nutrition.pdf

23. Committee on Animal Nutrition, *Nutrient Requirements of Nonhuman Primates*, Second Revised Edition (Washington, DC: National Academies Press, 2003), pp. 137–149.

24. Gorilla (*Gorilla gorilla*). Ensembl (Cambridge). Freely accessible (2011) at http://uswest.ensembl.org/Gorilla_gorilla/Info/Index

25. Jerry A. Coyne, *Why Evolution Is True* (New York: Viking, 2009), p. 68.

26. Coyne, *Why Evolution Is True*, p. 69.

27. Jennifer F. Hughes, Helen Skaletsky, Tatyana Pyntikova, Tina A. Graves, Saskia K. M. van Daalen, Patrick J. Minx, Robert S. Fulton, Sean D. McGrath, Devin P. Locke, Cynthia Friedman, Barbara J. Trask, Elaine R. Mardis, Wesley C. Warren, Sjoerd Repping, Steve Rozen, Richard K. Wilson & David C. Page, "Chimpanzee and human Y chromosomes are remarkably divergent in structure and gene content," *Nature* 463 (2010): 536–539.

28. Don E. Wilson & DeeAnn M. Reeder (editors), *Mammal Species of the World: A Taxonomic and Geographic Reference*, Third Edition (Baltimore, MD: Johns Hopkins University Press, 2005). Freely accessible (2011) at http://www.bucknell.edu/msw3/

29. Yuriko Ohta & Morimitsu Nishikimi, "Random nucleotide substitutions in primate nonfunctional gene for L-gulono-gamma-lactone oxidase, the missing enzyme in L-ascorbic acid biosynthesis," *Biochimica et Biophysica Acta* 1472 (1999): 408–411.

30. Yoko Inai, Yuriko Ohta & Morimitsu Nishikimi, "The whole structure of the human nonfunctional L-gulono-gamma-lactone oxidase gene—the gene responsible for scurvy—and the evolution of repetitive sequences thereon," *Journal of Nutritional Science and Vitaminology* (Tokyo) 49 (2003): 315–319.

31. Evgeniy S. Balakirev & Francisco J. Ayala, "Pseudogenes: Are They 'Junk' or Functional DNA?" *Annual Review of Genetics* 37 (2003): 123–51.

32. Amit N. Khachane & Paul M. Harrison, "Assessing the genomic evidence for conserved transcribed pseudogenes under selection," *BMC Genomics* 10 (2009): 435. Freely accessible (2011) at http://www.biomedcentral.com/1471-2164/10/435

GLOSSARY

Adenine: One of the four bases in the nucleotides in DNA and RNA.

Alu **sequence:** A retrotransposon in the SINE family, so named because it was first identified with an enzyme from the bacterium *Arthrobacter luteus. Alus* are the most common SINEs in primates.

Amino acid: A molecule with an amine group (NH_2) at one end, a carboxyl group (COOH) at the other, and a side group that distinguishes it from other amino acids. Proteins consist of chains of amino acids in which the amine group of one combines with the carboxyl group of another. Twenty amino acids are known to be encoded by DNA, but others also occur in living things.

Ancient repetitive elements: A term that rarely occurs in the scientific literature but is used by Francis Collins to refer to repetitive DNA.

Central Dogma: As formulated by Francis Crick, the idea that information can be transferred from nucleic acids to other nucleic acids or to protein, but not from proteins to proteins or nucleic acids. It is sometimes stated as "DNA makes RNA makes protein makes us."

Centromere: A special region of a eukaryotic chromosome that connects the chromosome to other structures in the cell. Just before a cell divides, duplicated chromosomes are connected by their centromeres.

CENP: CENtromere Protein, one of the several proteins associated with centromeres. Centromeres in all organisms depend on CENP-A, which takes the place of some of the histones in chromatin to provide a structural foundation for the centromere.

Chromatin: The combination of DNA, proteins and RNA that makes up a chromosome. It includes histones, special protein spools around which the DNA molecule is wound.

Chromosome: A microscopic thread-like structure in living cells that consists of chromatin.

Chromosome loop: A segment of chromatin that loops out from the body of the chromosome so that two distant parts of the DNA (such as an enhancer and promoter) can interact directly with each other at the ends of the loop.

Codon: A sequence of three adjacent nucleotides in DNA that specifies an amino acid in a protein or signals a ribosome to stop translation.

Cone cell: A photoreceptor cell in the retina that is involved in color perception and functions best in relatively bright light.

Conserved sequences: DNA or RNA sequences that are similar in different organisms. According to evolutionary theory, if two lineages diverge from a common ancestor that possesses DNA sequences that are nonfunctional, those sequences will accumulate mutations that render them different ("divergent") in the two descendant lineages. But if the original sequences are functional, then natural selection will tend to weed out mutations, and the corresponding sequences in the two descendant lineages will remain similar ("conserved").

Creationism: The religious view that the world was divinely created. In the modern controversies over Darwinian evolution it takes two general forms: young-Earth creationism and old-Earth creationism. After the first edition of *The Origin of Species*, Darwin added the statement that life was "originally breathed by the Creator into a few forms or into one," so broadly speaking Darwin might be called a creationist. But he believed that the evolution of living things after their initial creation could be explained without God's further involvement, and "creationist" is often used to describe someone who rejects this aspect of Darwin's view.

C-value paradox: Also known as the C-value enigma, this refers to the fact that the DNA content (the "C-value") of eukaryotic cells varies by a factor of several thousand, with no apparent correlation to organismal complexity or to the number of protein-coding segments ("genes").

Cytosine: One of the four bases in the nucleotides in DNA and RNA.

Dark matter: A term borrowed from physics, used in some junk DNA arguments to mean non-protein-coding DNA or RNA.

Darwinism: The theory of biological evolution according to which all living things have descended with modification from one or a few common ancestors by unguided processes—primarily random variations and natural selection. (As used in this book, "Darwinism" includes "neo-Darwinism.")

DNA: DeoxyriboNucleic Acid, which consists of nucleotides containing one of four bases (adenine, cytosine, guanine and thymine). In living cells, DNA occurs as a double helix composed of two complementary strands; during replication the two strands separate and serve as templates for the synthesis of new strands.

ENCODE Project: ENCyclopedia Of DNA Elements, a project of the U.S. National Institutes of Health to identify all the functional elements in the human genome.

Enhancer: A relatively short region of DNA that can increase the transcription of an open reading frame, which can be tens of thousands of nucleotides away on the same chromosome or even on a different chromosome.

Enzyme: A protein catalyst that increases the speed of a chemical reaction that would otherwise take place very slowly. Like an inorganic catalyst, an enzyme is not consumed by the process in which it participates.

Epigenetic: Etymologically, "above the gene." This adjective describes heritable changes in gene expression or the phenotype that do not involve changes in the nucleotide sequence of DNA. The change in chromatin that occurs when CENP-A replaces normal histones to provide a foundation for a centromere is epigenetic.

ERV: Endogenous RetroVirus, a genomic sequence that resembles (and might be derived from) the sequence of an RNA virus that has been reverse transcribed into DNA.

Euchromatin: A loosely packed form of chromatin, rich in protein-coding DNA sequences.

Eukaryote: A cell with a membrane-bound nucleus that contains the chromosomes, as in animals and plants.

Evolution: Etymologically, "unrolling." Originally used to describe the process of embryo development; later used to describe the history of the cosmos, living things, or human culture. Evolution can mean simply "change over time," which is uncontroversial. In biology it can also mean minor changes within existing species ("microevolution") or large-scale changes in the history of life ("macroevolution"). Darwinism is a particular theory of macroevolution.

Exon: A protein-coding segment of an open reading frame in DNA. Exons remain in the messenger RNA after introns have been removed, and if the RNA is translated into protein the exons specify the amino acid sequence.

FANTOM Consortium: Functional ANnoTation Of the Mammalian Genome, a project of the Japanese Riken Center that—like the U.S. ENCODE Project—is dedicated to identifying all the functional elements in the human genome.

Gene: Originally, an abstraction denoting the carrier of a Mendelian trait; later, the part of a chromosome carrying a Mendelian trait; later still, a segment of DNA (an "open reading frame") that encodes the amino acid sequence of a protein. In light of recent discoveries about the complexity of the genome and the transcriptome, the gene concept is now recognized to be an over-simplification.

Gene expression: The process in which the DNA sequence of an open reading frame encodes the synthesis of an RNA and/or protein. Expression can be regulated to produce varying amounts of the resulting RNA or protein.

Genome: Commonly used to mean the entirety of an organism's DNA, including the non-protein-coding portions.

Genotype: The set of an organism's genes—the protein-coding regions of its DNA. To be distinguished from "phenotype," the organism's anatomy, physiology and behavior.

GLO: L-GulonoLactone Oxidase (also abbreviated GULO), an enzyme that catalyzes the last step in the biosynthesis of ascorbic acid (vitamin C).

Guanine: One of the four bases in the nucleotides in DNA and RNA.

Heterochromatin: A tightly packed form of chromatin, poor in protein-coding DNA sequences but rich in non-protein-coding DNA.

Histones: Special proteins in the nuclei of eukaryotic cells that serve as spools around which DNA is wound in chromatin.

Homology: Originally, similarity of the structure and position of anatomical features in different organisms (such as bones in the forelimbs of vertebrates). Pre-Darwin biologists attributed homology to construction on a common design, but Darwin attributed it to inheritance from a common ancestor. Darwin's followers re-defined homology to mean similarity due to common ancestry, but the original meaning is still used, leading to ambiguity. The ambiguity persists in modern molecular biology, where homology can mean both similarity of nucleotide or amino acid sequence and similarity due to common ancestry.

Initiation sequence: A DNA sequence that signals the beginning of an open reading frame, where RNA polymerase starts transcribing DNA into RNA.

ID: Intelligent design, the idea that it is possible to infer from evidence in nature that some features of the world and/or living things are better explained by an intelligent cause than by unguided natural processes. Though often confused with them, ID is not the same as creationism or natural theology.

Intron: A non-protein-coding segment of an open reading frame in DNA. Introns are transcribed into RNA but removed before the RNA is translated into protein—though they contain codes that affect alternative splicing of the exons.

Inverted nucleus: A nucleus in which heterochromatin (normally located at the periphery) is concentrated in the center. The centrally located heterochromatin in the rod cells of a nocturnal animal serves as a lens to focus scarce rays of light.

Jumping gene: A segment of DNA that can move from one place to another in the genome. Such mobile genetic elements are called "transposons."

Junk DNA: DNA that is thought to perform no function in a living cell. People who assume that the only essential function of DNA is to code for proteins regard non-protein-coding DNA (about 98% of the human genome) as junk.

Kinetochore: A complex molecular structure that forms on a centromere and participates actively in moving chromosomes to the daughter cells during the process of cell division.

LINE: Long Interspersed Nuclear Element, a retrotransposon and one type of repetitive DNA. LINEs tend to be more than 5,000 nucleotides long and include a DNA sequence encoding an enzyme that enables them to reinsert themselves into DNA. Mammalian genomes contain tens of thousands of LINEs that fall into several groups; the most common of these is called L1.

LTR: Long Terminal Repeat, a sequence that flanks an endogenous retrovirus and is repeated hundreds or thousands of times.

Macroevolution: Large-scale changes in living things, such as the origin of new species, organs and body plans.

Mendelian genetics: The theory proposed by Gregor Mendel that the features of living things are determined by discrete heritable factors that were later called "genes."

Microevolution: Minor changes within existing species.

Microtubules: Microscopic tubules within eukaryotic cells that serve as structural supports and tracks for intracellular transport. Microtubules also move chromosomes during cell division.

Natural theology: A discipline that infers the existence and attributes of God from evidence in nature. Not to be confused with creationism or intelligent design.

NCSE: National Center for Science Education, a California-based organization dedicated (in its own words) "to keeping evolution in the science classroom and creationism out." By "evolution," the NCSE means Darwinism, and by "creationism," it means intelligent design as well as all forms of creationism. The NCSE also opposes the inclusion of evidence-based criticisms of Darwinian theory in science classes.

Neocentromere: An extra centromere that forms abnormally, either elsewhere on a chromosome that already has one, or on a chromosome fragment that has separated from the part bearing a normal centromere.

Neo-Darwinism: Darwinian theory combined with Mendelian and molecular genetics. Mendelian traits are carried by "genes" that program embryo development, and genes are equated with DNA sequences. Natural selection produces changes in gene frequencies (i.e., the relative proportions of variant DNA sequences), and new variations originate through genetic mutations (i.e., changes in DNA sequences due to replication errors or recombination).

Nucleotide: A subunit of the nucleic acids DNA and RNA. DNA consists of nucleotides containing the four bases adenine (A), thymine (T), cytosine (C), and guanine (G). RNA contains A, C, and G, but uracil (U) takes the place of thymine (T). RNAs may also contain other nucleotides in addition to these four.

Open reading frame: A segment of DNA that can be transcribed into RNA. All protein-coding genes are open reading frames—but not all open reading frames are genes, since their RNAs might not be translated into proteins.

Onion test: A challenge posed by biologist T. Ryan Gregory to anyone who proposes a universal function for non-protein-coding DNA. The challenge is to explain why an onion cell has five times as much non-protein-coding DNA as a human cell—an example of the C-value paradox.

Paraspeckle: A compartment in the nucleus that functions in gene regulation and is dependent for its stability on non-protein-coding RNAs.

Phenotype: The observable characteristics of an organism, including its development, anatomy, physiology and behavior.

Poly-A tail: A long tail attached to some RNAs that consists of many repeats of the nucleotide containing adenine (A) and is involved in the stability and translation of the RNA.

Primate: An omnivorous mammal with inward-closing fingers, fingernails, opposable thumbs, and a relatively large brain, belonging to a biological order that includes lemurs, monkeys, apes and humans.

Prokaryote: A cell without a membrane-bound nucleus, as in bacteria.

Promoter: A DNA sequence that provides a site for the attachment of RNA polymerase, which can then transcribe the nearby DNA into RNA.

Protein: A molecule consisting of a linear chain of amino acids that folds into a characteristic three-dimensional shape.

Pseudogene: A non-protein-coding segment of DNA with a nucleotide sequence that resembles a DNA segment that codes for protein elsewhere in the same organism or in other organisms. *Disabled (or unitary) pseudogenes* are single sequences that may have once coded for protein but have been inactivated by nucleotide changes or deletions. *Duplicated pseudogenes* are copies of still-functioning genes, though unlike the functioning originals they have characteristics that prevent them from encoding proteins. *Processed pseudogenes* have sequences similar to those of functioning genes, except that they lack promoter sequences and are usually missing introns.

Repetitive DNA: A DNA sequence that is repeated in the genome—in some cases, thousands of times. About half of the human genome consists of repetitive DNA, and about two-thirds of those repetitive sequences are LINEs or SINEs.

Replication: the process in which the two complementary strands of DNA separate and serve as templates for the synthesis of two more complementary strands; the result is two double-stranded DNAs that (barring mutations) have identical nucleotide sequences.

Retrotransposon: A mobile genetic element (transposon) that uses RNA as an intermediate in what amounts to a "copy and paste" process. The DNA element is first transcribed into RNA, then an enzyme called reverse transcriptase copies the RNA sequence back into DNA that is inserted into a different place in the genome.

Reverse transcription: A process in which the nucleotide sequence in a strand of RNA is copied into DNA; catalyzed by an enzyme called reverse transcriptase.

Ribonucleoprotein: A combination of one or more RNAs and proteins, such as a ribosome. Other ribonucleoproteins occur elsewhere in the cell, such as paraspeckles.

Ribosome: A large complex assemblage of RNAs and proteins that translates the nucleotide sequence of an RNA molecule into an amino acid sequence in a protein.

RNA: RiboNucleic Acid, which consists of four principal nucleotides (adenine, cytosine, guanine and uracil) and a number of less common nucleotides. RNA is normally single-stranded; some RNAs serve as templates for protein synthesis, but most RNAs perform a variety of other functions in the cell.

RNA interference: A process in which a non-protein-coding RNA reduces the expression of a gene by binding to—and thereby inactivating—the protein-coding RNA derived from that gene. RNA interference is one type of RNA silencing.

RNA polymerase: A large enzyme that synthesizes RNA with a sequence that matches a DNA template in the process of transcription.

RNA silencing: A process in which a non-protein-coding RNA reduces the expression of a gene. The most common form of RNA silencing is RNA interference, but silencing can also occur through the action of RNA-induced silencing complexes that cut up RNAs that might otherwise be translated into proteins.

Rod cell: A cell in the retina that is more sensitive to light than a cone cell and functions mainly in peripheral and night vision.

Satellite DNA: A fraction of DNA consisting of millions of short, repeated nucleotide sequences that produce "satellite" bands when DNA is centrifuged to separate it into fractions with different densities. Every normal human centromere is located on satellite DNA.

Selfish DNA: Junk DNA that appears to serve no other function than its own survival and persists as a parasite in its host cell.

Sequence Hypothesis: As formulated by Francis Crick, the idea that the specificity of a segment of DNA is expressed solely by the sequence of bases, and this sequence is a simple code for the amino acid sequence of a particular protein.

SINE: Short Interspersed Nuclear Element, a retrotransposon and one type of repetitive DNA. SINEs tend to be less than 500 nucleotides long and depend on other mobile genetic elements for their retrotransposition. The most common SINEs in primates are called *Alus*; rodent genomes contain different SINEs called B1 and B2.

Splicing: The process in which the exons in an RNA transcript are put back together after the introns are cut out. In alternative splicing, some exons may be omitted while others may be duplicated.

Syncytin: A protein derived from an endogenous retrovirus that plays an essential role in placenta development by contributing to the fusion of trophoblasts.

Tandem repeat: A form of repetitive DNA in which (usually short) sequences of nucleotides are repeated adjacent to each other. Satellite DNA consists of tandem repeats.

Target mimicry: A phenomenon in which a non-protein-coding RNA increases the expression of a gene by taking the place of that gene's protein-coding RNA in the process of RNA degradation.

Telomere: A segment of repetitive DNA at the end of a chromosome that protects the latter from degradation.

Termination sequence: A DNA sequence that signals the end of an open reading frame and stops transcription into RNA.

Thymine: One of the four bases in the nucleotides in DNA; in RNA it is replaced by uracil (U).

Transcription: A process that uses a DNA sequence as a template to synthesize ("transcribe") an RNA molecule ("transcript") with a matching sequence—except that uracil takes the place of thymine in the RNA.

Transcriptome: The entirety of an organism's RNA.

Translation: The process by which a ribosome converts the nucleotide sequence of a messenger RNA into the amino acid sequence of a protein.

Transposon: A mobile genetic element (known colloquially as a "jumping gene") that can move from one place in the genome to another, in what amounts to a "cut and paste" process.

Trophoblasts: Cells that are derived from a mammalian embryo and form a layer around it but are not incorporated into the fetus. Instead, they become part of the placenta, which supplies nutrients to the embryo and serves as the interface between it and the mother. In order for the placenta to function properly, some trophoblast cells must fuse into one giant, multinucleated cell (a "syncytium") called a "syncytiotrophoblast."

Uracil: The base in a nucleotide that takes the place of thymine in RNA.

X & Y chromosomes: Sex-determining chromosomes. In most mammals, each egg carries an X chromosome while each sperm carries either an X or a Y. If the egg is fertilized by a sperm carrying a Y chromosome the offspring is male (XY); if the egg is fertilized by a sperm carrying an X chromosome the offspring is female (XX). In order for the female to develop normally, one of its two X chromosomes must be inactivated.

INDEX

CPSIA information can be obtained at www.ICGtesting.com
Printed in the USA
LVOW061643041011

249072LV00002B/36/P